TOWARD A PSYCHOLOGY OF THE SCIENTIST

Implications Of Psychological Research For Contemporary Philosophy Of Science

Sonja C. Grover

UNIVERSITY
PRESS OF
AMERICA

P.O. Box 19101, Washington, DC 20036

Printed in the United States of America

ISBN (Perfect): 0-8191-1575-4
ISBN (Cloth): 0-8191-1574-6
LCN: 80-6092

To my parents and Roshan

Acknowledgements

I wish to thank Dr. Ian Lubek of The University of Guelph for having read the entire manuscript. His critiques and suggestions have been extremely valuable. To Dr. William Fowler of Harvard University I express my appreciation for his great encouragement of my academic endeavours.

I am deeply grateful to my parents and to my husband, Roshan. They have sustained me with their optimism concerning this project. A special thanks to my father for having proofread the manuscript.

Finally, I acknowledge the influence of the many who, through their works, have stimulated my thinking. In particular I should like to mention Dr. P. Feyerabend, Dr. T. Kuhn and Dr. M. Mahoney. I am appreciative also to those authors whose views about science are not in accord with mine for they, too, have been influential in my intellectual development.

CONTENTS

PREFACE

In this text, I have taken the liberty of relating studies from widely diverse areas of psychological research in order to make certain points. I have drawn from certain studies implications regarding questions to which these studies were not addressed originally. I can but hope that the authors of the studies mentioned will be appreciative of the fact that their work has led to speculation beyond those questions they had in mind initially, and that they find this text somewhat insightful.

This text is an attempt to contribute to what Mahoney(1) terms the "psychology of the scientist". Mahoney's urging for investigations into the psychological processes of the scientist is reminiscent of Bannister's statement, made approximately a decade before the Mahoney text was published, that psychologists ought to theorize about their own scientific activities. Bannister writes that "we have not yet faced up to the issue of reflexivity and the need for reflexivity in psychological thinking" (2, p.5). While Bannister held that scientists in fields such as physics and chemistry need not necessarily speculate upon the psychological factors influencing their work, Mahoney's text presents cogent arguments to the contrary.

It should be noted, given the argument in favor of reflexivity here advanced, that the author's model of scientific behavior is subject to all the limitations in describing "reality" that are all other models discussed in the pages that follow. That is, it may be quite possible to challenge the author's model by reinterpretation of supportive data cited, attention to additional evidence overlooked and so on. Such an eventuality would, however, itself serve as an example of the processes the author's model attempts to describe. The model suggests that scientific progress occurs via reformulations of arbitrary world views. Such views are often of great practical relevance to be sure, as evidenced by the technological advancement which are their by-product. Yet these models remain but imaginative constructions. As Eccles puts it: "Science is . . . a way of looking at things . . . a way of imagining" (3, p.141).

An attempt has been made to remedy what Fisch (4) regards as a major methodological weakness in studies of the psychology of the scientist to date - the failure to compare the psychological processes of the scientist to that of the nonscientist. Hopefully, some important parallels have been drawn by this author between cognitive process in the scientist and nonscientist.

While most of the research examples used in this text are drawn from the psychological literature, this is not to imply that the particular issues and problems which are discussed are peculiar to psychological investigations. Rather, these examples were employed since they are most familiar to the author. To discover the psychological and social-psychological factors which have defined what science has become is, it is suggested, a necessary step if there is to be a fuller understanding of the mechanisms underlying scientific progress.

This book would be of interest to any psychology student who wishes to learn more about the nature of psychological research, and the role of theory, imagination and creativity in experimentation generally. It would be of special interest to philosophy students, especially those concerned with issues in the philosophy of mind and philosophy of science. The discussions are relevant also to the issue of in what respects psychology, as all other sciences, is an experimental philosophy. An attempt is made to illustrate, via specific references and case studies of different lines of psychological research, how the philosophical assumptions of the scientist are in large part determiners of experimental findings. The theme of the book is that psychological research provides insights into how scientists come to formulate and revise theories. Illustrations are provided of how such psychological research furthers understanding of how scientific progress is made.

The book stresses the subjective side of the scientific endeavour, a side students are rarely formally introduced to via texts. Also for the psychology student, the book puts the issue of the scientific status of psychology in a new light by arguing that all sciences are in large part philosophical in nature. By illustrating, via specific examples, that theories are not simply tested against the "facts" or deduced from observations; a new appreciation for the tentative nature of human knowledge will, it is hoped, be encouraged as well as a greater respect for the role of imagination in stimulating scientific progress.

REFERENCES

1. MAHONEY, M. *The Scientist as Subject: The Psychological Imperative.* Cambridge: Ballinger, 1976.
2. BANNISTER, D. Psychology as an exercise in paradox. In Schultz, D.P. (ed.) *The Science of Psychology: Critical Reflections.* Englewood Cliffs: Prentice-Hall, 1970, p. 4-10.
3. ECCLES, J. *Facing Reality: Philosophical Adventures of a Brain Scientist.* London: Longmans, 1970.
4. FISCH, R. Psychology of Science. In Spiegel-Rösing, I. and De Solla Price, D. (eds.) *Science, Technology and Society,* London: Sage Publications, 1977, p. 277-318.

INTRODUCTION

ALTERNATIVE VIEWS OF THE NATURE OF SCIENTIFIC INQUIRY

The purpose of this text is to raise certain issues which hopefully will highlight for the psychological researcher (or potential researcher) alternative conceptions of the nature and function of experimentation in the psychological sciences. In addition, the author's objective is to elucidate the unique contribution which psychological research, particularly in the area of cognitive process, can make to an understanding of the scientific endeavour as a whole.

It will be argued in this text that psychology is fundamentally similar to all other sciences in its being an "experimental philosophy". The term "experimental philosophy" will appear to be a contradiction in terms only to those readers, it is contended, who do not fully appreciate the theoretical (philosophical) and speculative nature of empirical evidence. To view science as an "experimental philosophy" is to blur the distinction between theory and data but it is not, as will be discussed, to eliminate the demarcation between science and pseudo-science. This work is then not in any sense to be interpreted as "anti-science" in orientation.

The notion will be put forth that erroneous claims have been made regarding the knowledge which arises as the result of the scientific enterprise. It will be suggested that the data which arise as the result of scientific experimentation are a reflection of the scientists' mental constructions or conceptions rather than a reflection or manifestation of nature in the raw. The question thus arises as to why, if such data are but a reflection of the researcher's thoughts and not a true "picture of reality," scientific theories should ever "work;" predictions should ever be confirmed. These questions concerning the correspondence between theory and data will be addressed. It will be suggested that the correspondence is not so startling, given the fact that theory and data are not separable, but are determined by one another and that both are in turn the products of the scientist's imaginative processes.

Such distinguished researchers as Ulric Neisser have referred to psychology as an "unusual humanity" in that it is "one that collects data" (1, p.162). It is here contended, however, that psychology is no more nor no less closely allied to the humanities than are other scientific disciplines in that all sciences presuppose certain theories of how humans function, process data, and what their values are, and so are similar in this respect to humanities. The distinction between the so-called "hard core" sciences such as physics, chemistry and "soft-core" sciences such as psychology will, it is hoped, become less meaningful as the result of consideration of the points raised.

Illustrations will be given of how various lines of psychological research have clarified, to a degree, the nature of human knowledge. Psychological research has revealed something of how information is selected and organized and the inevitable distortions which result. It has uncovered some of the unconscious theories upon which the scientist and nonscientist alike rely in interpreting data, whether the data be the result of "objective" experimentation or less systematic methods of collecting empirical evidence.

The insights which psychological research can provide into what permits scientific progress to occur will also be considered. Scientific progress, it will be suggested, results when anomalous data are perceived which contravene accepted theoretical formulations. The author is in accord with Feyerabend (2) who holds that in order to be able to perceive evidence which conflicts with currently held conceptions, or at least in order to attach any significance to it, requires that the scientist be able to entertain novel, alternative views. These views would be in opposition, at least partially, to those prevalent at the time. Psychological research, for instance in social psychology and cognition, provides clues as to which factors create resistance to the adoption of alternative theoretical views and which make for a more flexible attitude. Research in social psychology concerning the notion of "locus of control," the degree to which one perceives oneself to be in control of one's destiny, may provide insights into the origin and development of certain philosophical assumptions underlying scientific inquiry. For example, the notion that one can potentially be in control of one's destiny - in that the events which one experiences are not random but are predictable and hence more than likely open to manipulation - can be examined in the context of such research.

To suggest that psychology has something to offer in terms of adding to an understanding of what facilitates scientific progress is to enter into a long-standing debate between opposing philosophical parties. Karl Popper (and I paraphrase him) argues that it is:

> a foolhardy idea to turn for enlightenment on the issue of the aims of science and its possible progress to psychology, for . . . psychology is often a spurious science (3, p.57-58).

Others, such as Thomas Kuhn (4), hold quite the opposite position, and suggest that it is necessary to study social psychological processes as they operate within the scientific community if the mechanisms underlying scientific processes are to be comprehended. By reviewing more specific evidence from psychological research, it is hoped the reader will be in a better position to judge just what contributions such work can make to an understanding of how new theories evolve and the factors which determine their fate; their acceptance or rejection.

Scientific inquiry and progress is generally held to be a monument or testimony to the fruitfulness of rational thought. Psychological research, however, reveals something of the degree to which irrational conceptions are essential for creativity and hence for scientific progress. By the term "irrational" is meant that which is meaningless within a certain context or conceptual framework. Many ingenious notions appear "irrational" or "senseless" when viewed from old perspectives, but later are found to be scientifically productive. The question of the role of "irrational" concepts in stimulating scientific progress then will be discussed. An attempt will be made to learn from the psychological literature what processes are involved when certain irrational notions come to be regarded as legitimate and gradually become part of the contemporary "world view."

It is frequently not the case that novel theories become accepted by the scientific community because they are shown to correspond with the (empirical) facts. Theories are often accepted though they are at present untestable. Such theories refer, for instance, to ideal states which do not exist such as the ideal gases referred to in Boyle's law. The example of Boyle's law is given by Hebb who states that:

> science is a way of taking off into the wild blue yonder, and it gets harder and harder to understand how it stays in touch with . . . reality (5,p.10).

It is here suggested that the psychological literature provides insights into how it is that the "preposterous ideas" of physicists (and other scientists), to use Hebb's terminology, manage to stay in contact with reality; that is how it is that they appear to give accurate descriptions of the world.

The role of imaginative processes in science, as Hebb (6) so eloquently outlines, has been underestimated. Perhaps this is because the distinctive feature of science has traditionally been thought to be its empirical nature; its reliance upon testable theories and the method of controlled experimentation. Kuhn (7) points out that testable theories are not the feature which distinguishes science from pseudo-science. Astrologists, Kuhn explains, also propose testable theories and make predictions which are open to disconfirmation. Rather than the testability of theories, Kuhn argues, it is the nature of the research puzzles which the scientist invents that sets him apart from his nonscientific fellows. The research puzzles which the scientist poses during certain periods when paradigms clash challenge currently held assumptions. They involve a leap of the imagination beyond what is accepted. While the astrologer resists posing puzzles which threaten the very tenets upon which astrological tradition is formulated, the scientist glories in proposing alternative views as is evident in the following quote from the great experimentalist Michael Faraday:

> The laws of nature, as we understand them, are the foundation of our knowledge in natural things . . . they have become, as it were, our belief or trust . . . We have no interest in their retention if erroneous; on the contrary, the greatest discovery a man could make would be to prove that one of these accepted laws was erroneous, and his greatest honour would be the discovery (8, p.463).

That is not to say that there is never psychological resistance within the scientific community to new theories. Despite empirical evidence that attests to the superiority of the new model or theory, the new view may not win acceptance easily. For instance, Bohr's model was retained despite evidence to the contrary as was Newton's gravitational theory. As Brush puts it:

> scientists often operate in a subjective way . . . experimental verification is of secondary importance compared to philosophical arguments, at least in some of the major conceptual changes that have occurred in science (9, p.1166).

Often then theories are accepted or rejected not on the basis of the empirical "evidence" but upon the basis of the persuasibility of the new theoretical argument itself. Psychology has much to say on the issue of what makes theoretical arguments psychologically persuasive, and how novel theoretical arguments come to be proposed.

The cognitive psychology of the scientist has long been a neglected topic, perhaps in large part because the role of cognitive processes in scientific research has been oversimplified. The scientist is generally conceptualized as making logically necessary inductions (generalizations) on the basis of observations, and then making logical, testable deductions from hypotheses. However, inductions from observations and deductions from theory are far from being logically necessary or dictated by the data. Brush describes scientific paradigms as:

> including not only a theory, but also a set of criteria for determining what problems are worth solving and how one recognizes a solution when he has it (10, p.1167).

Theories are "self-policing", to use an analogy, they specify what the rules of evidence shall be, and what constitutes a confirmation or disconfirmation of the theory. For instance, operant researchers contend that removing an effective treatment variable results in unsatisfactory behaviour by the subject similar to that occurring during baseline. Reintroduction of the variable, such as a particular reinforcement schedule, produces the appropriate behaviour once more thus illustrating its control over behaviour. This argument is advanced despite maintenance of the appropriate behaviour at follow-up; again in the absence of the treatment variable (11, p.356). The model thus specifies which criteria for theoretical adequacy apply, and when. Such a process is not without its inconsistencies as the above example illustrates. With any self-policing, then, there is the possibility for a definite bias. Psychological research in the areas of, for instance, perception and cognition reveals much about how such biases operate and how unaware one can be of them.

To state that it is possible to adduce evidence for any theory is not so wild a claim if it is realized that in point of fact no theory is firmly tied to empirical evidence. The link between hypothesis and data is a theoretical one and as such a cognitive one. Appeal to the method of falsification does not eliminate the problem for data do not intrinsically bear upon theoretical propositions for the purpose of disconfirmation either. Theories consist of a set of statements from which only further statements can be *logically* deduced. Observations cannot be so deduced. It is incorrect then to hold that any particular theoretical view logically implies a certain set of observations. The observations gain meaning in terms of the theory and hence are inextricably bound to it. It is, furthermore, the theorist and not the rules of logic who defines what ought to follow from the theory; what predictions it suggests. Hence, it is possible for different investigators to disagree as to what empirical predictions are the logical consequences of the theory. These deductions depend on how the researcher comprehends the theory, and the meaning of its terms. Here too, psychological research has important information to offer as to the nature of comprehension and the determinates of linguistic meaning.

This author is not the first to argue for psychological study of the scientific enterprise, (cf. 12, 13, 14, 15, 16), but it is hoped that in presenting specific examples from the psychological literature which bear upon certain fundamental epistemological issues and questions of scientific methodology, the point will be made more comprehensible as well as compelling. What is needed, it is argued, is a reconceptualization of the nature of scientific knowledge and a greater appreciation of the psychological factors involved in the formulation, and acceptance or rejection of world views (theories). It is hoped that this book may serve to clarify the point that in large part our psychology - the way in which our minds function - determines what we discover not only about ourselves but about the world.

Lakatos terms as revolutionary those scientists who contend that:

> conceptual frameworks can be developed and also replaced by new, better ones; it is we who create our prisons and we can also critically demolish them (17, p.104).

The ability to reconceptualize our own nature, its limits and possibilities, as well as to adopt new world views, it is held, will be immeasurably enhanced as we come to further comprehend our own psychological functioning.

REFERENCES

1. Neisser, U. Self-knowledge and psychological knowledge: Teaching psychology from the cognitive point of view. *Educational Psychologist,* 1975, *11,* p. 158-70.
2. Feyerabend, P.K. Against Method: Outline of an Anarchistic Theory of Knowledge. In Radner, M. and Winokur, S. (eds.) *Minnesota Studies in the Philosophy of Science,* Vol. IV, Minneapolis: University of Minnesota Press, 1970, p.17-130.
3. Popper, K. Normal science and its dangers. In Lakatos, I. and Musgrave, A. (eds.) *Criticism and the Growth of Knowledge.* Cambridge: Cambridge University Press, 1970, p. 49-58.
4. Kuhn, T. Logic of discovery or psychology of research? In Lakatos, and Musgrave (eds.), *Criticism and the Growth of Knowledge,* p. 1-23.
5. Hebb, D.O. Science and the world of imagination. *Canadian Psychological Review,* January 1975, Vol. 16, No. 1, p. 10.
6. Ibid.
7. Kuhn, *Logic of discovery or psychology of research?* p. 1-23.
8. Faraday, M. Observations on mental education. *Experimental Researches in Chemistry and Physics.* London: R. Taylor and W. Francis, 1859.
9. Brush, S.G. Should the history of science be rated X? *Science,* 1974, *183,* p. 1166.
10. Ibid., p. 1167.
11. Mahoney, M. Psychology of the scientist: An evaluative review. *Social Studies of Science,* 1979, Vol. 9, p. 349-375.
12. Kuhn, *Logic of discovery or psychology of research?* p. 1-23.
13. Mahoney, *Psychology of the scientist,* p. 349-375.
14. Bannister, D. Psychology as an exercise in paradox. In Schultz, D.P. (ed.) *The Science of Psychology: Critical Reflections.* Englewood Cliffs: Prentice-Hall, 1970, p. 4-10.
15. Fisch, R. Psychology of Science. In Spiegel-Rösing, I. and De Solla Price, D. (eds.) *Science, Technology and Society.* London: Sage Publications, 1977, p. 277-318.
16. Maslow, A.H. *The Psychology of Science.* New York: Harper and Row, 1966.
17. Lakatos, I. Falsification and the methodology of scientific research programmes. In Lakatos and Musgrave (eds.) *Criticism and the Growth of Knowledge,* p. 91-195.

CHAPTER I

ASPECTS OF CONTEMPORARY PHILOSOPHY OF SCIENCE

This section of the text attempts to elucidate the views of the contemporary philosopher of science, Thomas Kuhn and to demonstrate that Kuhn's position finds much support in the psychological literature. Three aspects of Kuhn's philosophy will be focused upon as they have had perhaps the greatest impact on contemporary philosophical thought: (a) his view that there exists no basic set of data independent of theory to which all theories are addressed; (b) his position that rival paradigms are incommensurable and (c) his view that there are never compelling reasons to favor one scientific paradigm over another. The choice, according to Kuhn, is an irrational one based on a conversion to an alternative world view (1, 2).

There are important similarities between Kuhn and other philosophers of science such as Feyerabend on these points in certain respects. Kuhn's work is used then as representative of certain philosophical themes which are prevalent in many contemporary works. Each of these three aspects of Kuhn's philosophy is discussed in what follows.

ON THE FATE OF "RAW DATA": DATA INDEPENDENT OF THEORETICAL FORMULATIONS

The contemporary philosopher of science is much more likely than would have been his predecessors to appreciate the subjective elements in scientific research. For instance, the hope of formulating a language of science devoid of bias and interpretive distortions, a language composed of "empirical statements" which described nature as it is has largely been abandoned. Such was the hope of the positivist philosophers of the Vienna Circle (3).

Contemporary philosophers of science argue against the positivist conceptions of "a universal, non-theoretical, permanent, empiricist language adequate to express and compare the observational content of all physical theories. . ." (4, p.48). Psychological research in the areas of perception and cognition would appear to provide support for the notion that a truly empirical language - one that describes what is; without some degree of distortion is an impossibility. This because data are perceived by human observers whose sensory processes are not simply mechanical sensing devices, but also interpretive ones influenced by certain conceptual schemes.

At every stage in the analysis of sensory input interpretation occurs with a concomitant possibility for distortion. For instance, in the initial registration of a sensory input; it is characterized as "new" or "old" (familiar). If the stimulus input is "old", there will be less of an orienting response to it than if it were novel. The mechanisms involved may be either conscious or unconscious. The brain then monitors the input in terms of past experience and modifies the organism's responsiveness to incoming stimulation partly on the basis of such an analysis. To recognize a stimulus input as old or familiar may not be the result simply of searching memory for evidence of previous exposure to the stimulus in question; but may instead involve complex inferential analyses which sometimes involve errors. This is demonstrated in recent memory studies. These demonstrate that both children and adults tend to misperceive novel sentences, presented during a sentence recognition task, as having been previously presented during sentence acquisition if these novel sentences represent true, logical inferences that could be deduced from the original acquisition sentences (5, 6).

What is "new" and what is "old" is thus not simply a matter of frequency of exposure, but is a function also of the brain's interpretive processing. It appears that the brain responds as much to the informational value of the stimulation the world has to offer as to the physical parameters of the stimulation. The paradox generated by the latter point is that the informational value of the stimulation is largely a byproduct of cognitive processes and not inherent in the input inself. For instance, in listening to speech the hearer derives information from the perceived speech segments and their combinations. Yet these speech segments are not in the physical input, the sound wave, which is but a continuous stream without segmentation. The information seemingly derived from the speech signal is in fact in large part superimposed upon it. Stimulation does possess structure, but such structure is apparent and informative only to the perceiver who analyzes and organizes the input in the appropriate or requisite manner. Hence, for example, the word units in a particular speech sample are perceptible only to the listener familiar with the language spoken.

An empirical, scientific language which describes raw data is not possible then because the data which it would describe will have already been distorted to a degree by human perceptual and conceptual processes. There is no body of "raw" data to be described since data are not only immediately interpreted by the perceiver, but its structure is in fact generated by particular cognitive frameworks such as the theoretical framework of the scientist. According to Kuhn (7) not only does a new scientific paradigm assign new importance to certain sets of data, but it allows also for observations which would have been impossible to make given the concepts of the older theoretical view. Data then are generated by or are a function of particular conceptual perspectives. The structure of the data is also a product of conceptual processes. To describe such structure is to express a certain world view and not simply to offer an objective description of what is.

Not all psychologists would hold to the views expressed. Most notable of the opponents to the position just put forth would probably be J.J. Gibson (8, 9). It is the view of Gibson that stimulation is inherently structured and that perception of this structure can provide direct knowledge of environmental properties (10). For instance, Gibson explains that as an object recedes from view, its textural elements appear to become more and more dense thus providing a cue for distance of the object from the observer. Gibson's model provides a useful account of how such cues are used to order our sensory experience. Gibson concedes that perception of linguistic input is a product in part of expectation (conceptual analysis of the input) and not just a product of the stimulation provided by the speech signal itself (11). He contends, however, that this is not the rule for perception in general.

The evidence that the structure of the linguistic input is in no simple sense "in" the input itself is abundantly clear from the early research conducted by Huey (12). Huey demonstrated that readers often perceived as complete, words with letters missing when reading these from a distance. Furthermore, their subjective certainty in perceiving the absent letters was often times greater than their certainty in having perceived letters which were in fact printed. Subjects in Huey's studies had filled in the missing elements in the word structure and could not distinguish the structure provided by the orthographic regularities in the letters printed from that invented structure resulting from the imaging of absent letters on the basis of expectations as to what word was printed.

It is contended here that both the structure of the imaged letters and that of the letters actually printed were constructions, and emerged as the result of certain expectations. The structure would not be present prior to the formulation of such expectations. It is a product of the cognitive processes which are involved in making the input meaningful. Structure and meaning are in fact inseparable as evidenced by Pritchard's (13) finding that when the stable images, of either pictorial or linguistic stimuli, projected onto the retina fade, they do so in terms of meaningful units. The image does not fade as a set of single, unrelated lines. For instance, the word "heart" may fade in such a way that first the "he" component disappears followed by the "art" component. Meaning (in this case the word meaning which identifies "he" and "art" as two subcomponents of the larger image) defines what are the structural components of the image. There are an unlimited number of ways in which the lines which form the word configuration "heart" could be organized. That the image fades by first the "he" and then the "art" component disappearing is a reflection of an arbitrary organization (structuring) of the input into particular subcomponents. The structure "in" the stimulus input (in the retinal image projected by the word "heart") is a product of the conceptual processes which assigned this organization to the input, it is not intrinsic to the visual configuration itself.

The Gestalt reversible figures reveal that the structure of input is a conceptual construction for nonlinguistic input also. It is not possible to perceive both figures at once as each requires a different structuring of the input. The brain alternates between two mutually exclusive interpretations. Once having structured the input in one way, however, it becomes difficult to call up the alternative interpretation; for this requires the application of a new meaning (e.g. the young woman versus the old hag). Yet with attentiveness to the input and a concentration on what one hopes to find there, the alternative figure emerges from the input. But does this occur simply because the data here are ambiguous? This author contends all input is in fact ambiguous until structured via conceptual schemas.

Semantic satiation effects also reveal the constructive aspects of perception. Over-repetition of a word results in the input being heard as new words. There is a degree of unrecognized flexibility in how the input can be interpreted or "structured". To contend, as does Gibson, that stimulation is inherently structured, and ready to be assimilated in a structured form is, it is suggested, to begin the analysis of perceptual processes at too late a stage. It is to overlook the considerable cognitive work which has preceded this point, and which underlies the perceiver's ability to apprehend the input as organized or structured in a particular fashion.

Gibson's view is consistent with the positivist view that data can exist independently of any particular set of theoretical assumptions. Data, in the positivist view, can be conceptualized as inherently patterned or structured stimulations of scientific import. It is here suggested, in opposition to such a view, that just as linguistic perception is influenced by expectation and cognitive frameworks, so too, is non-linguistic perception.

Many skilled scientists recognize, to a degree, the role of the perceiver in modifying the structure of different types of stimulation. They appreciate, unlike proponents of a view of science as completely objective, that data generated by experiments are not completely independent of theory. Mitroff found that a view of theory and data as inseparable was not foreign to the Apollo moon scientists, for instance. He cites one of these men as saying:

> I don't believe the observation of raw data is ever really independent of theory. The two things constantly interact (14,p.183).

At the same time, however, Mitroff found there was no clear acceptance by his subjects that data are an extension of theory; a different formulation of the same information. Considerable confusion arose in response to the interview statement:

> "Theories of science are not tested by data which are arrived at independently of the theory which is being tested."

Sample responses to the above item included:

> I'm not sure what you mean here. My answer is mixed . . . I'd agree in the sense that it may not be possible to test the most fundamental theories of science by data which is not somehow dependent on the theories being tested. But I believe this is possible with the more routine, smaller theories.

Another scientist responded to the same item as follows:

> Theories should be tested by data arrived at independently of the theory being tested (15, p.183-184).

There appears then often to be some degree of conflict experienced by scientists who have been trained in the orthodox view that theory stimulates experiments that yield unrelenting data which then serve as an independent test of the theory. First hand experience in the scientific process most frequently does not fit such a model.

In what follows examples are drawn from physics, astronomy, and psychological research on developmental process and the intellect to illustrate how data (structured stimulation) in a sense emerge and are defined by theory.

CASE EXAMPLES OF CONSTRUCTIVE PROCESSES IN PERCEPTION

The N-ray affair

The problem to which Gibson does not address himself is why it is that cues function as cues, and how they come to be noticed and regarded as significant. It is here suggested that Gibson's model does not necessitate that the structure be "in" the stimulation. That structure could as well be the product of the perceiver's selectivity, his expectations and attentional processes which organize the input in certain ways in order to make sense of it. Cues become cues, it is here contended, via such active structuring and transformation of the sensory data. It is only, furthermore, when a particular construction of cues is challenged that it is apparent that the structure was imposed rather than perceived directly.

Consider the discovery of the N-Ray; now thought to be non-existent. The N-ray was "discovered" by Blondlot, a reputable researcher in the physics of electromagnetic radiation. Considerable empirical evidence was gathered to substantiate the existence of the new ray and to demonstrate its properties. Blondlot's studies were replicated by many other investigators of the French Academy of Sciences. Not until some twenty years after the discovery of the N-ray was its existence successfully challenged by non-believers who devised alternative experimental designs and criteria for the test. These studies demonstrated that the previous empirical cues which had led to the inference about N-rays were unreliable (16).

Blondlot's case is not unique in that the evidential base in science is constantly being revised and the significance of particular sets of data as cues for certain inferences being reinterpreted. The discarded model or theory is not so much a "mistake", but rather reflects a particular construction of data which loses its importance as new cues previously not apparent are attended to (cf. 17). Those new cues become evident due to the novel conceptions of scientists bent on challenging prevailing models and accounting for their inconsistencies. These cues (patterns) are then not simply "in" the data.

A remark by James Clerk Maxwell, founder of the mathematical theory of electromagnetism, is relevant to the points raised. Maxwell was speaking about the competing theories of light and noted:

> There are two theories of light, the corpuscle theory and the wave theory; we used to believe in the corpuscle theory; now we believe in the wave theory because all those who believed in the corpuscle theory have died (18, p.175).

In other words, the significance of particular data patterns is very much dependent upon those scientists who perceive these as cues for the making of particular theoretical assumptions and inferences.

The Apollo Scientists

Mitroff's study of the Apollo moon scientists revealed that they had shifted their views very little on hypotheses relating to, for instance, the moon's origin and geological composition after the mission. The lunar data collected seemed to have little immediate impact upon hypothesis selection (19, p.158). The significance of the lunar data, its cue function, for many of these scientists had then not yet been constructed.

Developmental Studies: Language Development, Hypothesis Formation, Intellectual Process

That the patterns or structure in the data and the data itself become evident only after certain theoretical formulations have been made is brought out forcefully in a beautiful study by Karmiloff-Smith and Inhelder (20). These authors observed children of four to nine attempting to balance blocks of various dimensions and weights across a narrow bar. Some of the blocks (length blocks) were so constructed that they balanced at their geometric centre. Others were constructed in a way that resulted in their balancing across the bar only at points outside their geometric centre. The latter blocks either had weights attached which were clearly visible (conspicuous weight blocks), or weights concealed within each end (inconspicuous weight blocks). Younger children had little difficulty balancing any of the three types of blocks across the narrow bar inasmuch as they relied upon proprioceptive feedback. At around six to seven and one-half years, the children appeared to have formulated the theory that all blocks balance at their geometric centres, and inappropriately overgeneralized this theory to blocks with weights attached.

Of great importance was the children's tenacity in retaining this "geometric centre theory" despite the empirical evidence to the contrary. This evidence being that the blocks with weights attached failed to balance at their geometric centres. With eyes closed these children could once again rely on proprioceptive feedback and balance all the blocks. With eyes open, however, the counterexamples provided by the weight blocks were not taken to be counterexamples since the child's conceptual framework did not allow for such contradictions of the theory. Only gradually did children beyond the age of seven come to modify the "geometric centre theory". First they came to realize that the conspicuous weight blocks balanced at points other than their geometric centres, and only later did they recognize that the same was true for nonconspicuous weight blocks.

The data which are attended to, the pattern suggested by such data and its relevance to theory is then dictated by the theory itself. The children in the Karmiloff-Smith and Inhelder study could not take account of the counterexamples to their theory because they were not cognitively prepared to attach significance to the contradictory empirical data. A similar phenomenon is evident in the research of scientists. For instance, Donald Hebb speaks of the tendency of psychologists "of coming to conclusions by believing theory rather than looking at people" (21, p.15). Hebb notes that until recently psychologists held that intelligence declined from the age of twenty onward. This despite the numerous examples of persons making significant intellectual contributions in middle and old age. Such data were conveniently not attended to; though this evidence would appear on the surface to have been quite accessible.

Another example of failure to take note of what would appear to be patently obvious comes from the research on language development. It had until quite recently been held that language acquisition was primarily the result of an exact imitation and correction type of process. Such a model fails to incorporate explanations for the inventiveness of children's early speech. Children, as is evident to parents, fabricate many idiosyncratic verbal expressions which could not have resulted from imitation. Also neglected was the evidence that there are large individual differences between young children in the degree of imitation of adult linguistic inputs - another readily available piece of data.

It would appear that data achieve significance and are attended to only if the theory permits. For instance, not until cognitive theories were advanced was the creativity of early childhood speech considered to be significant data of theoretical interest. It should be clear then that raw data independent of theoretical formulation are for the reasons just discussed but a myth, as is the possibility of a strictly objective language which describes such raw data. Yet this myth persists for it is at least the ideal of many scientists to use data as an *independent* test of theory (22).

Incommensurable Theories: Theories of Hypnosis as a Case in Point

According to Kuhn, different scientific paradigms are not directly comparable insofar as each has its own set of problems to deal with and a distinct range of data considered to be relevant to those problems. There is considerable controversy concerning the notion of paradigm, "Kuhn held it to include law, theory, application and instrumentation together" (23). To avoid such controversies competing theories rather than paradigms will be examined here, yet Kuhn's point is equally applicable in this instance.

Two contrasting theories of hypnosis are to be discussed. The first views hypnosis as an altered state of consciousness and the second, the theory of Theodore Barber (24) conceives of hypnosis not as an altered state, but merely as a state of heightened suggestibility. This example is appropriate in that it is quite unclear whether in fact Barber's studies offer an alternative interpretation or explanation for so-called hypnotic effects or not. The problem arises in that both theories appear plausible and are supported by confirmational data. Yet what "counts" as confirmational data for either of the perspectives is a matter of controversy, and depends not so much upon objective assessment of the "empirical facts" as upon one's faith or lack of it vis-à-vis the existence of the phenomenon termed "hypnosis".

Barber terms his view a cognitive-behavioral one, an approach which "proceeds to account for so-called "hypnotic" experiences and behaviors without postulating that the subjects are in a special state of hypnotic trance" (25, p.5). In Barber's view, "subjects carry out so-called hypnotic behaviors when they have positive attitudes, motivations and expectations toward the test situation which lead to a willingness to think and imagine with the themes suggested" (26, p.5). Barber's research has led to the identification of numerous factors which enhance an individual's susceptibility to instructions. For instance, simply labelling the situation as one involving hypnosis is sufficient to enhance responsiveness to test suggestions. A critical feature of the situation then appears to be that it leads the subject to define the situation as one of hypnosis. Barber holds that relaxation-sleep-hypnosis suggestions raise responsiveness to test suggestions, not because they induce an hypnotic trance, but because they "define the situation to the subject as "truly hypnosis" - a situation in which high responsiveness to suggestions is desired and expected and in which they should try to think and imagine with the themes suggested" (27, p.25-26).

The question could be raised however as to whether or not readiness to think and imagine with the themes suggested is not a sufficient criterion for determining that the individual is in an hypnotic trance. The trance is not similar to that of a somnabulist, at least along the particular dimensions studied, as evidenced by the "fact" that the EEG of a sleepwalker is quite dissimilar from that of an hypnotized subject (28). This state of extremely high suggestibility, however, might yet be characterized as a trance in that the subject may not be in conscious control of all of that behavior of which he normally is. Barber himself suggests "If the subject responds to suggestions and judges from his responses that he is hypnotized, this might heighten his expectancy that he will respond to further suggestions and the heightened expectancy may influence his subsequent response" (29, p.44).

It could be argued then that the hypnotist has created in the subject false expectations as to his degree of self control vis-à-vis the test instructions. This may be something other than simply a state of heightened motivation to comply with test instructions. It may involve a fundamental misattribution of the causes of behavior. The subject perceives his behavior as being externally controlled via test suggestions. Spiegel (30) describes the hypnotist as "tacitly implying (to the subject) that the phenomena are due to the signal of the hypnotist". For instance, the hypnotist, Barber explains, might ask the subject to stand with his eyes closed which for everyone results in a slight body sway. The hypnotist notes the direction and rhythm of these natural body sways and times his suggestions of sway with these. The subject erroneously comes to believe his body sway is a consequence of the hypnotist's suggestions and gradually the sway in a particular direction becomes more pronounced.

It could be argued that insofar as these subjects genuinely accept their misattributions of causes as correct, they are not simply role-playing at being hypnotized. To present evidence that role-playing subjects not exposed to hypnosis can emulate much of the behavior of "hypnotized subjects" does not in fact provide direct evidence that such effects are always simply the result of role-play. Indeed, some experiments seem to suggest certain differences between subjects who are role-playing and those who believe themselves to actually be hypnotized. For instance, while hypnotized subjects experienced certain colour contrast effects due to imaging certain colours on darker or lighter backgrounds, such effects were not found for role-playing subjects (31). The hypnotized subjects were differentially susceptible to test suggestions compared to the role playing subjects.

The author has no preference for either the hypnotic trance model or the cognitive-behavioral interpretation of hypnosis advanced by Barber. Rather the point of this discussion has been to illustrate that competing theories do not attach the same significance to particular bodies of data. The public nature of science hence does not assure that the soundest models persist. What is considered confirmational data from one theoretical viewpoint is considerably less convincing from another theoretical vantage point. The controversy thus rages on as to what all the data mean, providing further insight not only into the nature of hypnosis but the nature of scientific theorizing itself and its relation to the accumulation of data. To choose one theory over the other, as Kuhn has pointed out, is not as rational a process as is generally claimed. It appears that most often in science, particular theories are chosen from among competitors first, and the data achieve their convincing confirmational nature as a consequence of the selection.

REFERENCES

1. Kuhn, T. *The Structure of Scientific Revolutions,* International Encyclopedia of Unified Science, Vol. 12, No. 2, (2nd edition enlarged), Chicago: University of Chicago, 1970.
2. Kuhn, T. Logic of discovery or psychology of research? In Lakatos, I. and Musgrave A. (eds.) *Criticism and the Growth of Knowledge.* Cambridge: Cambridge University Press, 1970, p. 1-23.
3. Kolakowski, L. *The Alienation of Reason: A History of Positivist Thought.* New York: Anchor Books, 1968.
4. Doppelt, G. Kuhn's Epistemological Relativism: An Interpretation and Defense, *Inquiry,* 1978, Vol. 21, No. 1, p. 48.
5. Paris, S.G. and Carter, A.Y. Semantic and Constructive Aspects of Sentence Memory in Children. *Developmental Psychology,* 1973, Vol. 9, No. 1, p. 109-113.
6. Brewer, W.F. Memory for the Pragmatic Implications of Sentences. *Memory and Cognition,* 1977, Vol. 5, No. 6, p. 673-678.
7. Kuhn, *The Structure of Scientific Revolutions.*
8. Gibson, J.J. Perception as a Function of Stimulation. In Koch, S. (ed.), *Psychology: A Study of a Science.* Vol. 1, New York: McGraw-Hill, 1959, p. 456-501.
9. Gibson, J.J. *The Senses Considered as Perceptual Systems.* Boston: Houghton & Mifflin, 1966.
10. Mace, W.M. Ecologically Stimulating Cognitive Psychology: Gibsonian Perspectives. In Weimer, W.B. and Palermo, D.S. (eds.) *Cognition and the Symbolic Processes.* New York: Erlbaum Associates, 1974, p. 137-164.
11. Turvey, M.T. Constructive Theory, Perceptual Systems, and Tacit Knowledge. In Weimer, W.B. and Palermo, D.S. (eds.) *Cognition and the Symbolic Processes,* p. 165-180.

12. Huey, E. *Psychology and Pedagogy of Reading.* Massachusetts: M.I.T. Press, 1968 (1st ed., 1908).

13. Pritchard, R.M. Stabilized Changes on the Retina. *Scientific American,* 1961, 204, p. 72-77.

14. Mitroff, I.I. *The Subjective Side of Science: A Philosophical Inquiry into the Psychology of the Apollo Moon Scientists.* New York: Elsevier, 1974.

15. Ibid., p. 183-184.

16. Klotz, I.M. The N-Ray affair. *Scientific American,* May 1980, Vol. 242, p. 168-175.

17. Kuhn, *The Structure of Scientific Revolutions.*

18. Klotz, *The N-ray Affair,* p. 175.

19. Mitroff, *The Subjective Side of Science,* p. 158.

20. Karmiloff-Smith, A. and Inhelder, B. If you want to get ahead, get a theory. *Cognition*, 1975, 3(3), p. 195-212.

21. Hebb, D.O. "On Watching Myself Get Old". *Psychology Today,* November, 1978, p. 15. (Article adapted from an invited address at APA, Toronto, 1978).

22. Mitroff, *The Subjective Side of Science,* p. 183-184.

23. Shapere, D. The Paradigm Concept. *Science,* May, 1971, Vol. 172, p. 706-709.

24. Barber, T., Spanos, N.P. & Chaves, J.F. *Hypnosis, Imagination, and Human Potentialities.* New York: Pergamon Press, 1974.

25. Ibid., p. 5.

26. Ibid.

27. Ibid., p. 25-26.

28. Ibid.

29. Ibid., p. 44.

30. Spiegel, H. Hypnosis and Transference. *Archives of General Psychiatry,* 1959, 1, p. 634-639.

31. Zimbardo, P.; *Psychology and Life.* Diamond Printing, Ninth ed. Glenview: Scott Foresman and Co., 1977.

CHAPTER II

PLAYING BY THE RULES OF THE SCIENCE GAME

METHODOLOGICAL RULES: PSYCHOLOGICAL ASPECTS OF EXPERIMENTAL DESIGN

Bannister several years ago made the amusing but apt observation that:

> Had Christopher Columbus . . . possessed the mind of many modern psychologists, I am reasonably certain he would never have discovered America. To begin with, he would never have sailed because there was nothing in the literature to indicate that anything awaited him except the edge of the world. Even if he had sailed, he would have set forth bearing with him the hypothesis that he was travelling to India. On having his hypothesis disconfirmed when America loomed on the horizon he would have discovered the whole experiment null and void and gone back home in disgust (1, p.6).

It is here suggested that there is a certain logic in experimental design that coincides with a particular "psycho-logic" which is not particularly helpful, and which makes Bannister's comment more accurate than one would hope it were. It would appear that there are certain biases built into experimental design and data interpretation which lead the scientist often to retain theories despite disconfirming evidence and to count as confirmational, evidence (data) which in fact may not necessarily be so. The end result of these biases, most often unconsciously held, is that scientists find it difficult, as do nonscientists, to alter in any substantial way their conceptual framework.

ILLUSIONS OF CONTROL AND CONFIRMATIONAL BIASES

Experimental design is such that it allows for control of extraneous variables and isolation of causal variables or so at least is the general conception of the nature of adequate research designs. What is argued here is that experimental design often leads to a mis-attribution of causal factors precisely because it has become an à priori assumption that design in and of itself can "separate out" factors. Experimental design allows for the manipulation of the independent variables of interest held to be tampered with by the scientist independently of other potentially relevant factors. When effects are obtained with extraneous variables controlled, it is assumed the independent variable manipulation was assuredly responsible for the same. While such a deduction is well-accepted, it would seem to be faulty. The plausibility of the notion stems from the fact that such thinking characte-rizes the causal analysis of most adults whether they be scientists or not. Kelly states:

> The purpose of causal analysis - the function it serves for the species and the individual - is effective control . . . Controllable factors will have a high salience as candidates for causal explanation. In cases of ambiguity or doubt, the causal analysis will be biased in its outcome toward controllable factors (2, p.22-23).

It appears to be the case then that experimental design, with its manipulation of independent variables, is likely to lead the scientist to attribute causal responsibility for any effects produced to the independent variable simply because the variable was manipulated.

At once, the reader will no doubt argue that such will be the case only if the effects produced clearly confirm the hypothesis that specified a certain independent variable as being critical. Yet the psychological literature, and historical examples unfortunately present a more complicated picture than that. It seems that persons are prepared to attribute causality to manipulated or controllable factors if these are attended to and expected to be causal, regardless of overwhelming disconfirmational data. The N-ray "discovery" mentioned in the previous chapter illustrates this point. Despite a wealth of disconfirming evidence, Blondlot insisted upon the reality of his finding and the adequacy of his experimental manipulations.

Conversely, factors not manipulated may be regarded as noninfluential, and even escape observation. Thus, for example, the discovery of x-rays, which violated certain expectations, was resisted in part because the "earlier scientists had failed to recognize and control (this) relevant variable". The scientist, Lord Kelvin, who initially proclaimed the findings a hoax, was not prepared to concede the causal influence of these rays which must have been produced by others besides Röntgen "without knowing it" (3, p.59).

Several psychological studies also illustrate such a process. A study by Jenkins and Ward (4) and another by DeMonbreun and Mahoney (5) are examined next in that these are fairly representative of current lines of research concerned with problem-solving, causal attribution and its relation to confidence in hypotheses.

In the Jenkins and Ward study, experimental subjects were instructed to work on five problems. Outcomes were experimentally varied and subjects were told that the relationship between their response (a button press) and the outcome (score or no score) could range from no control to complete control. After sixty trials on each problem, subjects were asked to rate on a numerical scale their perceived degree of control over the outcome. The results revealed that subjects consistently misperceived their degree of control. The perceived degree of control was positively related to the frequency with which the subjects produced the "score" outcome. Subjects tended to infer that being able to produce an intended outcome implied a great deal of personal control over outcomes even when intended or desired outcomes occurred infrequently. For instance, on one of the problems subjects received only eight successes in sixty trials and there was in fact no contingency between their responses and the desired outcome. Yet one-half of the subjects concluded they had some control over the outcome. It appears that "If a person attempts to produce a certain outcome and that outcome occurs he is very likely to ignore the possibility that the outcome occurred by chance" (6,p.31).

Studies by Mahoney (7) reveal that scientists too are subject to such confirmational biases. Scientists have been found to make the logical error of affirming the consequent - assuming that if a conclusion is true, the premise must also be so. This bias is manifested in their scientific research in terms of the tendency to assume that confirmational findings necessarily attest to the validity of an hypothesis. Affirming the consequent is illogical, and misleading in that "false" theories can and often do lead to "correct" predictions.

It would appear that manipulated variables are perceived as being causal and that there is therefore often a distortion in terms of how data are assessed. So powerful is the tendency to perceive oneself as in control of an experimental situation that negative data are frequently perceived to be invalid, for instance, due to flaws in the experimental design rather than due to erroneous theorizing. These psychological factors, it is suggested, underly in large part the reluctance of scientific theorists to revise their perspectives. As Kuhn (8) points out, the history of science reveals a certain resistance to changes in world view. Such alterations in world view come about usually as the result of clashes between opposing paradigms. Rarely does the change emerge as the result of an "inside job", it is rather more commonly a question of "break and enter" from the outside.

THE SCIENTIST'S RELUCTANCE TO CHANGE WORLD VIEWS

Barber (9) reviews a number of instances of resistance of scientists to important scientific discoveries. These include resistance by Helmholtz to Max Planck's notions about the second law of thermodynamics, opposition by Von Nägeli to Mendel's theory of genetic inheritance, and rejection of Röntgen's discovery of the X-ray by Lord Kelvin, to mention but a few. So disheartened was Max Planck about the reception his notions received that he stated he "found no interest let alone approval . . ." (10) for his novel conceptions among his colleagues.

A more recent example is provided by the experience of Ito, a neurologist who worked with John Eccles. Ito, with his own research group in Tokyo, later discovered that the whole output of the cerebellar cortex is inhibitory. Eccles writes candidly that Ito's findings:

> were so original that I just rejected them . . . It is important to appreciate that inevitably we have tensions between innovations and the orthodoxy of the Establishment . . . Ito . . . discovered something quite unexpected, fundamental and new, and my reaction had immediately been to reject it, and say that something was the matter with his measurement of time, or that his electrodes were not properly recording, or something like that, because evidently I belonged to the establishment . . . (11, p.142-143.

Eccles did eventually come to accept Ito's findings. Acceptance of novel ideas in science is then usually hard won. Often, as Barber explains, new discoveries are initially discounted as they cannot be easily accommodated by prevailing models (12).

Resistance to change can occur "both between two or more scientists but also within an individual scientist" (13, p.598). Thus, for instance, scientists first ignored the significance of their own anomalous findings that a particular enzyme injected into a rabbit's ears resulted in floppiness, for it demanded a new view of the nature of cartilage (14).

Such a tendency to discount negative or anomalous data comes out clearly in a study by DeMonbreun and Mahoney (15). In this study, subjects were asked to formulate an hypothesis as to what rule generated a series of numbers. For instance if the numbers were 17, 34, 68; 2, 4, 8; 1, 2, 4 one might hypothesize that the rule was: "multiply each successive number by two." Subjects were to rate their confidence in their hypothesized rule from 0% to 100% confidence. Subjects were then presented with new sets of numbers and instructed to decide whether the rule applied to this new set of numbers or not. After making their decision on each trial, subjects were given experimental feedback as to whether their decisions were correct or incorrect. Subjects were informed that 20% of the feedback was in fact false. After receiving the experimenter feedback, subjects were to rate anew their confidence in their hypothesized rule.

Experimenter feedback was so arranged that subjects either received early positive feedback, early negative feedback or random feedback. In each case, 50% of the feedback was negative. Subjects demonstrated a consistent confirmational bias. Regardless of the pattern of the experimenter feedback, subjects tended to perceive positive feedback as valid and negative feedback as invalid. They also maintained high levels of confidence in their original hypothesis with few shifts in confidence throughout the study, despite the fact that 50% of the experimenter feedback suggested their hypothesis was false.

The tendency to attribute causality to those factors which appear to be under control is not mitigated by the nature of the findings, confirmational or disconfirmational. There appears to be then not only a predilection for misattribution of causality, but also a concomitant misperception of the validity or lack of validity of the data in question. Given the fact that it is theory which specifies what are the relevant data and the significance of data, the muddle over just what implications the data have for any theory is futher exacerbated. The controversy over whether or not "experimenter bias effects" are substantially affecting psychological research provides a prime example of such a muddle and is discussed in this chapter.

It appears that people are prone to making certain basic errors in information processing which have important implications for the practice of science insofar as scientists are not immune to the same types of errors. Scientists, as nonscientists, would appear to be susceptible to the "illusion of control" in that both often appear to attribute causality to manipulated variables despite evidence to the contrary. The history of science provides many examples of scientists tenaciously holding on to particular theoretical frameworks despite negative results; of researchers "adjusting" data rather than modifying theory, and of selectivity in terms of what data are attended to, namely that data that appear to confirm the theory (16, 17, 18): This historical evidence leads a well known contemporary researcher, Zimbardo, to conclude:

> It is almost never the case that a theory is overthrown and discarded by facts to the contrary. Instead, worthless theories persist with a life of their own until replaced by others proven to be less objectionable (19, p.22).

The scientist, as Mahoney notes, seeks often to confirm his theory rather than to disconfirm it. In the search for viable paradigms, experimental design and statistical analyses are often used as means of legitimizing preexisting biases and not as means of eliminating them. Confirmational biases are clearly evidenced in the tendency of psychological researchers to perceive as valid, confirmational findings whenever the significance level is acceptable, regardless of the strength of the relationship between the variables at issue (20). Levine points out that:

> . . . the fact that only minute proportions of variance are accounted for by experimental variables, is not typically embarrassing to the experimenter . . . The evidence on p values, rather than on estimates of variance accounted for by treatments, indicates we are more concerned with socially meaningful decision criteria than with the power of the variables we manipulate (21, p.668).

Even when the statistical results are not in the theory's favor, this does not necessarily represent a deathblow to the preferred theory (22) for the data are often selected and organized so as to appear supportive. Furthermore, "experimental outcome is [often] used as evidence for the validity of vaguely stated verbal conceptual hypotheses" (23, p.668), where the link between experimental outcome and theory is particularly unclear. It is precisely because theorists define the significance of data and because the bearing of data upon theory is itself a matter of interpretation that rarely are theories forgone for the mere sake of disconfirmational data. Rather, it is due to "techniques of persuasion", according to Feyerabend (24) that new world views are adopted. Thus philosophical arguments which define what evidence there is, if in fact such exists, are sufficient for adoption of the new conceptual framework.

To sum up the first part of the discussion of the psychological aspects of experimental design, I will quote a beautiful phrase of John Dollard's: "The basic research method is the human intelligence trying to make sense of what it observes" (25). It is clear that experimental design which supports certain biases such as misattribution of causality is often a useful tool, but not a guide for the discernment of reality. Current models of experimental design clearly encourage confirmational biases and illusions of control since they are based on the assumption that theory can be completely distinguished from data, that the latter, given an appropriate experimental situation, can provide an independent test of the former. Current experimental thinking then creates the illusion that data are products of experimental manipulations; that they simply emerge as nature is unveiled via application of the objective scientific approach. It is here suggested, in contrast, that data are very much the product of imagination rather than the end result of objective manipulation and measurement techniques.

Levine states: "Given the assumption that the basic research instrument is the human intelligence, it follows that controls . . . should be those that take into account forms of bias to which the human intelligence is subject" (26, p.665). Hopefully, this book will provide a helpful overview of some of the psychological literature which provides insights into those "forms of bias to which the human intelligence is subject." Certainly, one type of control against bias would be afforded simply by the knowledge that scientific research is actually a matter of philosophizing about the meaning of data, the data itself being an expression of theory. Empirical tests of theory are but a form of argumentation, it is contended, their empirical nature should not mask the role of interpretation in the testing of theory. Gregory put the matter much more elegantly when he stated: "Scientific observations without hypotheses are surely as powerless as an eye without a brain's ability to relate data to possible realities - effectively blind" (27, p.707). Empirical tests are tests only to the degree that theory specifies their relevance. Scientific observations in and of themselves never challenge theoretical perspectives. It is rather philosophical argument which attaches a particular significance to observations (data) that transforms observations into tests which may challenge particular perspectives.

At this point, it is relevant to examine some of the social psychological literature which provides further insights into the reluctance to shift world views which often characterizes scientist and nonscientist alike. The area to be examined is locus of control research.

Locus of Control

The scientist, it has been pointed out in the preceding discussion, is subject to illusions of control. This illusion is made possible in part by a fundamental belief in determinism. It is here contended however that scientists could function quite adequately if they disregarded the notion of determinism and that in fact this notion underlies, to a degree, the misplaced faith that scientists often place in confirmational data. Confirmational data are taken to be evidence of a stable world revealing its regularities, and it becomes difficult then to challenge the significance of the data.

What the evidence from locus of control research seems to indicate is that persons prefer to view their world as stable and predictable; a world in which most things can be accounted for rather directly. There is a preference for the simplistic explanation as opposed to the complex. These tendencies would not appear to be entirely useful in that they tend to mitigate imaginative reinterpretations of what is and what might be and present therefore a special hindrance to the researcher.

It will be agreed by most, that scientists are persons with a fundamental faith in their ability to manipulate the environment in order to facilitate the making of predictions, and hence enhance control over a multitude of factors. While engaged in scientific activity, it may be assumed that the experimenter experiences a high sense of control insofar as it is the experimenter who manipulates the independent variables, controls extraneous variables, and defines and delimits an experimental situation in such a way that only certain events could possibly emerge in that experimentally defined context and not others. There would appear to be both positive and important negative consequences arising from the scientist's high degree of internal locus of control. The benefits of such an orientation include the fact that it leads persons to become more adequate information processors. For instance, in a study by Lefcourt and Wine (28) it was found that subjects who were internals were more likely to attend to cues which help them resolve uncertainties. Lefcourt states: "Internals have been found to be more perceptive to and ready to learn about their surroundings. They are more inquisitive, curious and efficient processors of information than are externals" (29, p.65).

The possible disadvantage of an extreme internal locus of control orientation would seem to be a reluctance to alter one's views under certain circumstances when a modification in outlook is called for. Lefcourt in reviewing evidence from particular studies on this issue concludes: "When the stakes of success are of some value to the individual, persons characterized as internals are more trusting of their own judgments than are externals . . . externals have more confidence in the consensual judgments of others than they do in their own independent judgments" (30, p.41).

It may be that the image of the scientist who would readily modify a theory in which he has invested much time and energy in reaction to the judgments of other scientists is but an idealized one. It is an open question as to how freely individual scientists accept criticism of their work and there are of course variations among researchers in this regard. Certainly, the history of science would seem to suggest that eminent scientists not infrequently prefer to hold on to failing theories despite attacks from colleagues until their position is totally untenable. Such a tendency to resist influence regardless of whether it derives from an authoritative source may in fact characterize persons with high internal locus of control (31). The desire to maintain one's theoretical stance leads inevitably to certain types of biasing. Such well-defined perspectives (biases) may, however, be critical to the scientific process in that they allow weaknesses of particular models to be highlighted.

THE SCIENTIST'S CAUSAL ATTRIBUTIONS

There are striking parallels between the scientist and nonscientist with regard to the distortions in data assessment which they are prone to make, some of which have already been noted. Research in the area of cognitive social psychology provides important additional insights into the mechanisms underlying such effects. In any psychological experiment, it is here suggested, there are at least two types of analyses of data performed. One set of analyses is formal and explicit and is represented by standard statistical procedures; the other is informal and implicit but yet no less critical to the final interpretation of the results. Attribution research has included an examination of this implicit analysis of data which most of us carry out with little conscious awareness. The point of interest for the purpose of the present discussion is the finding that such implicit analyses of data often lead to unsubstantiated conclusions. The question arises then as to whether scientists in engaging in similar informal, causal analyses bias the more formal statistical findings of their studies and how such biasing comes about.

The implicit nonformalized analysis of data gives rise to certain causal attributions which often times, as was mentioned previously, are inaccurate. Consider first such a process in the nonscientist. Jones et al. (32) demonstrated that persons often misattribute the causes of their behaviour. In their study, subjects provided with error feedback about their performance on a "rigged" I.Q. test attributed their own performance to external factors - task difficulty - while they attributed others' performance to variations in ability on the basis of identical error patterns.

It would appear possible that researchers, too, might often misattribute the causes of their (scientific) behavior. A significant misattribution would involve misattributing their assigning of a particular interpretation to a piece of data to external factors, the pattern in the data, rather than to internal factors. These internal factors might include their own preconceptions as to what the data are and the desire for theoretical simplicity and consistency. Thus the scientist assumes he has controlled the variables in the experimental situation in order to permit patterns "in" the data to emerge. He does not, however, generally view himself as having created those patterns by delimiting the experimental situation, or as having imposed interpretations upon data which are arbitrary.

Might not such a psychological process account for the Principia's "pretense to a degree of precision quite beyond its legitimate claim?" (33). It was the more precise calculations presented in the Principia which Newton presumed to be more valid than those he must have actually arrived at initially on the basis of experimental data. Newton probably attributed his judging of data not to any internal motivations but rather to the exigencies dictated by the "experimental facts" as he intuited them. Brush presents an additional example of faulty causal attribution by the scientist. He discusses the case of the predominance of one force law over another, not due to its superior fit with experimental data but its better match with theoretical calculations. Of particular note is the fact that the law which in fact had the poorest fit with the experimental data "continued to be described and used as the *most realistic* function in many works on statistical mechanics for . . . 30 years, despite the absence of any experimental basis for this claim" (34, p.1169). It is here suggested that the scientists who accepted the less accurate law, misattributed the causes of their behavior in selecting this law rather than the available more experimentally precise alternative. They described the force law selected as "more realistic" perhaps because they attributed their support for it to external causes - the experimental data - rather than to internal causes such as the desire for theoretical simplicity which the law provided.

What types of causal attributions the scientist makes in accounting for the reasons underlying his theoretical and quantitative findings are integral then to the experimental process and need to be scrutinized. It is fallacious to assume that theory simply derives from data, and that objective quantitative analyses of the data allows one to judge in any simple, direct fashion the validity of theory. It is also necessary to examine the psychological factors underlying causal attribution by the scientist in order to discover what factors gave rise to the relationship between data and theory which he perceives.

PSYCHOLOGICAL FACTORS INFLUENCING THE PERCEPTION OF THE RELATION BETWEEN DATA AND THEORY: A CASE IN POINT

An example from the psychological literature of differences in causal attribution between theorists, and how such differences lead to biases in the assessment of data is provided by the controversy between Rosenthal and Barber and Silver regarding the extent of the experimenter bias effect. The experimenter bias effect involves the tendency of experimenters to inadvertently or intentionally influence the outcome of experiments.

In Rosenthal's view the "effect may well be a fairly general one" (35, p.310) and is clearly demonstrated in a number of studies. In the view of Barber and Silver (36), the effect is not as widespread as Rosenthal claims. Barber and Silver review a number of studies which Rosenthal et al. claim support the notion of an experimenter bias effect. In the view of Barber and Silver, the studies cited by Rosenthal fall into two sets: "Those that failed to demonstrate an experimenter bias effect, and those that do not lend themselves to clear-cut conclusions because of one or more of the following inadequacies in the analysis of data:

1. An overall statistical analysis was not performed to exclude chance findings.
2. The authors performed a large numer of post hoc statistical tests after the overall analysis had failed to reject the null hypothesis at a conventional level of significance . . .
3. Problems of probability pyramiding were not avoided . . .
4. The authors strained for significance by accepting questionable p values (e.g. p values $< .10$) as substantiating the experimental hypothesis.
5. Negative data . . . were not used in the statistical analysis that ostensibly showed the experimenter bias effect . . ." (37, p.345).

Experimenter biasing effects are taken to include all the means by which an experimenter, though often in good faith, may bias his results. Means of mediating such an effect include the unintentional misrecording of data and unconscious selectivity as to which data are taken into account. Given such a broad view of the nature of experimenter biasing effects, it may be concluded that Barber and Silver are *unwittingly* arguing that Rosenthal's studies themselves fell prey to such biasing in that these authors allege Rosenthal's methodology had serious and obvious flaws, which were certainly avoidable, such as failure to use "no-expectancy" control groups. Hence, Barber and Silver place themselves in the awkward position of pointing out the types of methodological errors in Rosenthal's studies which imply experimenter bias, while at the same time arguing that his studies were not executed adequately enough to provide any support for the notion of experimenter bias.

The question arises as to why Barber and Silver did not perceive Rosenthal's allegedly inadequate statistical analyses and the other methodological flaws in his studies as in themselves support for the Rosenthal effect. Their failure to do so could be considered yet another example of experimenter bias in that Barber and Silver were not conceptually prepared to attribute the cause of such inadequacies to such biasing factors, which they reject for the most part. The issue here is *not* whether Barber and Silver are correct or not in finding fault with Rosenthal's studies. Rather, the point is that these authors were prepared to make certain causal attributions and not others, it is contended, regardless of the data. Whether or not Rosenthal's studies support the notion of experimenter bias is very much a function then not only of experimental results and statistical analyses, but of implicit and often misleading causal attributions which bear a very loose relationship with the data at hand.

Up to this point the rules defining acceptable research design have been considered. The scientific endeavour is guided, however, by a number of additional implicit and explicit rules. In the next section are examined implicit rules regarding ethical questions in scientific research and the larger issue of whether scientific skepticism is a hindrance to ethics.

ETHICAL RULES GUIDING RESEARCH

According to Polanyi: "As long as science remains the ideal of knowledge, and detachment the ideal of science, ethics cannot be secured from complete destruction by skeptical doubt" (38, p.27). Following is an elaboration of the views which appear to have led Polanyi to such a conclusion and an evaluation of his claim.

Scientists, Polanyi explains, generally believe that "the facts and values of science bear on a still unrevealed reality" (39, p.190). In his view, the "scientist can conceive problems and pursue his investigation only by believing in a hidden reality on which science bears" (40, p.192). He argues further that there is a widespread "failure to recognize man's capacity for anticipating the approach of hidden truth" (41, p.193). According to Polanyi, the orginality of the scientist is tempered by his felt responsibility to make contact with some external reality, hence "choices made in the course of scientific inquiry are . . . responsible choices made by the scientist, but the object of his pursuit is not of his own making" (42, p.194).

According to Polanyi then, science is an intensely personal endeavor involving commitment and fundamental metaphysical beliefs such as the belief that science can discover something of reality. Imagination, in his view, plays a critical role in the scientific process, but it is guided by intuitions about reality.

> We are able to know (in some anticipatory, intuitive sense) enough of what we do not know as yet in any explicit sense (because we have not yet discovered it) to enable us to locate a good problem and to begin to take groping but effective steps toward its solution (43, p.178).

This author conceives of scientific imagination as being unfettered by such intuitions of reality, in contrast to Polanyi, a point of contention taken up further in chapter five.

These passages from Polanyi are cited here to indicate the basis upon which he holds that the knowledge provided by science is fundamentally similar to the knowledge provided by the humanities. Specifically, he argues that these two sources of knowledge both involve imagination, firm beliefs rather than skepticism, and that neither in the case of humanistic nor scientific knowledge can "the grounds — metaphysical, logical, or empirical — upon which our faith that such knowledge is true be specified" (44, p.61).

This author is in full accord with Polanyi that scientific knowledge has no more favored a relation to reality than does humanistic knowledge. However, it is here contended that skepticism is not detrimental to ethics, and that the belief in the relation of certain ethical principles to some external truths is not a prerequisite for morally just behaviour. Morally just behaviour, it is suggested, is dependent upon a willingness to make choices in a moral dilemma. It derives from a sense of felt personal responsibility to act justly (45). This does not, it is held here, imply that morally just choices are dictated by universal ethical principles; contrary to the view held by Kohlberg who states: "true principles guide us to the obligating elements in the situation" (46, p.61).

It is here contended, in contrast to Polanyi, that just as the scientist is not directed in the making of his discoveries by intuitions of reality, so too, the individual in choosing that which appears to be just is not guided by intuitions of some universal moral principle such as the principle of justice. Consider, for instance, a situation in which justice would seem to dictate mutually exclusive lines of action. In such frequently occurring moral dilemmas, it is essential to remain skeptical about whatever choice is made in order to fully appreciate that apparently just behaviour may yet have unjust components. These issues have special importance to the researcher, and will be illuminated by a consideration of the ethical dilemma involved in the Stanford Prison Experiment conducted by Zimbardo.

Zimbardo conducted research on the question of how the prison environment — a situational factor — might affect an individual's personality, attitudes and values. Zimbardo used subjects who assumed the role of prisoner or guard for an extended period of time in a mock prison set up at the Stanford University Campus.

> The projected two-week stay had to be prematurely terminated when it became apparent that many of the 'prisoners' were in serious distress and many of the guards were behaving in ways which brutalized and degraded their fellow subjects (47, p.243).

The study served to illustrate just how forceful are the situational determinants of behavior. It underscored the fact that persons can be led to behave brutally, who would not have considered themselves capable of the same, merely as the result of being assigned a particular role which affords them certain powers.

This particular study generated much controversy. It is a striking example of the point made by Medawar that "The mischief science may do grows just as often out of trying to do good . . . as out of actions intended to be destructive" (48, p.82). The study provided important information about the determinants of immoral conduct and made a strong case for the need for prison reform. Nevertheless, the study exposed subjects to experiences involving suffering to which they normally would not have been exposed. The study then combines both just and unjust elements, and it is an open question as to whether it would have been more ethical for Zimbardo to have relinquished such a project or to have pursued it as he did.

The point could be raised that all scientific research contains beneficial and non-beneficial components. Most often these detrimental consequences of research are unforeseeable, Medawar (49) points out, as in the case of X-irradiation which was found to lead to cancer, and antibiotics which led eventually to an increasingly resistant strain of organisms which are immune to the effects of antibiotics. Scientific research then requires a deep sense of personal responsibility for the choices which are made in the final analysis are ethical ones. Skepticism is critical lest any person claim to have valid knowledge of what is just, leaving no room for doubt as to what is in fact the moral choice, and hence no opportunity for reappraisal of choices.

There is one additional issue which deserves mention at this point. According to Polanyi, science often involves a kind of mechanical reductionism. For instance, persons are conceived as organisms motivated by drives, or propelled into particular patterns of behaviour via conditioning histories or other types of variables. He fears that in this way science "may come to deny us personal responsibility" (50, p.25). In studies on compliance to authority or in the Zimbardo experiment, however, it is important to recognize that not all subjects are merely pawns given the situation — an element of choice remains. Hence in the Milgram study (51), where subjects were instructed to shock a victim (actually a confederate of the experimenter who did *not* in fact receive shock, though the subjects were unaware of this), not all subjects complied with the instructions of the experimenter to proceed in increasing the shock voltage every time the victim made an error in a learning task. Hence, even studies designed to demonstrate human behavior controlled by situational variables, also reveal the possibility for individual moral courage and personal responsibility.

Polanyi's latter point then is a critical one for scientific researchers to consider who would wish to establish harmony between their work and ethical principles. An encouraging development in psychological research is the increasing willingness to study the question of ethical behavior, a topic long thought to be quite outside the limits of legitimate scientific problems. Scientists yet need to become more self-reflective with regard to the ethical implications of their own behaviour as researchers.

REFERENCES

1. Bannister, D. Psychology as an Exercise in Paradox. In Schultz, D.P. (ed.) *The Science of Psychology: Critical Reflections*. Englewood Cliffs: Prentice-Hall, 1970, p. 4-10.
2. Kelly, H.H. *Attribution in Social Interaction*. Morristown: General Learning Press, 1971.
3. Kuhn, T. *The Structure of Scientific Revolutions*. 2nd. edition enlarged, Chicago: University of Chicago Press, 1970.
4. Jenkins, H.M. and Ward, W.C. Judgement of Contingency Between Responses and Outcomes. *Psychological Monographs*, 1965, 79 (Whole No. 594).
5. DeMonbreun, B.G. and Mahoney, M.J. The Effect of Data Return Patterns on Confidence in an Hypothesis. In Mahoney, M. *Scientist as Subject: The Psychological Imperative*. Cambridge: Ballinger, 1976, p. 181-186.
6. Wortman, C. Causal Attributions and Personal Control. In Harvey, J.H., Ickes, W.J. and Kipp, R.F. (eds.) *New Directions in Attribution Research*. New York: Lawrence Erlbaum Publishers, 1976, p. 23-52.

7. Mahoney, M. *Scientist as Subject: The Psychological Imperative,* 1976.

8. Kuhn, T. *The Structure of Scientific Revolutions.*

9. Barber, B. Resistance by Scientists to Scientific Discovery. *Science,* 1961, Vol. 134, p. 596-602.

10. Planck, M. *Scientific Autobiography.* Gaynor, F., trans. New York: Philosophical Library, 1949.

11. Eccles, J. *Facing Reality: Philosophical adventures of a brain scientist.* London: Longman, 1970.

12. Barber, *Resistence by Scientists to Scientific Discovery,* p. 598.

13. Ibid.

14. Ibid.

15. DeMonbreun, B.G. and Mahoney, M.J. The effect of data return patterns, p. 181-186.

16. Hebb, D.O. Science and the World of Imagination. *Canadian Psychological Review,* 1975, Vol. 16, No. 1, p. 4-11.

17. Brush, S.G. Should the History of Science Be Rated X? *Science,* 1974, Vol. 183, p. 1164-1172.

18. Barber, *Resistance of Scientists to Scientific Discovery,* p. 596-602.

19. Zimbardo, P.G. and Ruch, F.L. *Psychology and Life.* Diamond Printing, 9th Edition, Scott Foresman and Company, 1977.

20. Christensen, L.B. *Experimental Methodology.* Boston: Allyn and Bacon, 1977.

21. Levine, M. Scientific Method and the Adversary Model. *American Psychologist,* September, 1974, p. 661-716.

22. Westfall, R.S. Newton and the Fudge Factor. *Science,* 1973, Vol. 179, No. 4075, p. 751-758.

23. Levine, *Scientific Method and the Adversary Model,* p. 668.

24. Feyerabend, P. Classical Empiricism. In Butts, R.E. & Davis, J.W. (eds.) *The Methodological Heritage of Newton.* Toronto: University of Toronto Press, 1970, p. 150-170.

25. Dollard, J. Cited in Levine, *Scientific Method and the Adversary Model,* p. 665.

26. Levine, *Scientific Method and the Adversary Model,* p. 665.

27. Gregory, R.L. Seeing as thinking. An active theory of perception. *London Times Literary Supplement,* June 23, 1972, p. 707-708.

28. Lefcourt, H.M. and Wine, J. Internal versus external control of reinforcement and the deployment of attention in experimental situations. *Canadian Journal of Behavioural Science,* 1969, 1, p. 167-181.

29. Lefcourt, H.M. *Locus of Control: Current Trends in Theory and Research.* New York, John Wiley and Sons, 1976.

30. Ibid., p. 41.

31. Ibid.

32. Jones, E.E. and Nisbett, R.E. The Actor and the Observer: Divergent Perceptions of the Causes of Behaviour. In Jones, E.E. et.al. (eds.) *Attribution: Perceiving the Causes of Behaviour.* Morristown: General Learning Press, 1971, p. 79-94.

33. Westfall, *Newton and the Fudge Factor,* p. 751-758.

34. Brush, *Should the history of Science be Rated X?* p. 1164-1172.

35. Rosenthal, R. *Experimenter Effects in Behavioral Research.* New York: Appleton-Centruy Crofts, 1966.

36. Barber, T.X. and Silver, M.J. Fact, Fiction, and the Experimenter Bias Effect. In Miller, A.G. (ed.) *The Social Psychology of Psychological Research,* New York: Free Press, 1972, p. 342-385.

37. Ibid., p. 345.

38. Polanyi, M. *Meaning.* Chicago: University of Chicago Press, 1975.

39. Ibid., p. 190.

40. Ibid., p. 192.

41. Ibid., p. 193.

42. Ibid., p. 194.

43. Ibid., p. 178.

44. Ibid., p. 68.

45. Grover, S.C. An Examination of Kohlberg's Cognitive-Developmental Model of Morality, *Journal of Genetic Psychology,* 1980, Vol. 136, p. 137-143.

46. Kohlberg, L. Stages of moral development as a basis for moral education. In Beck, C.M., Crittenden, B.S. and Sullivan, E.B. (eds.), *Moral Education: An Interdisciplinary Approach.* New York: Newman Press, 1971, p. 23-92.

47. Zimbardo, P.G.; On the ethics of intervention in human psychological research: With special reference to the Stanford prison experiment. *Cognition,* 2(2), p. 243-256.

48. Medawar, P.B. *The Hope of Progress,* New York: Anchor Books, 1972.

49. Ibid.

50. Polanyi, *Meaning,* p. 25.

51. Milgram, S. *Obedience to Authority.* New York: Harper and Row, 1974.

CHAPTER III

THE ROLE OF PARADOX IN THE SCIENTIFIC RESEARCH PROCESS

According to Polanyi, who was both philosopher and scientist:

> it is thought to be useless to be puzzled by events which, although they may be explicable in principle, are not ripe for the explanation or are not worth the trouble of explaining them . . . in any case, a scientific explanation must serve to dispel puzzles (1, p.53).

In Polanyi's view then, scientific explanations serve to dispel puzzles. In this section, various paradoxes drawn from the psychological literature are examined. These include both paradoxes of a theoretical and/or an empirical nature. It is here contended that the ability to perceive and conceptualize paradoxes is an essential aspect of research skill and competence. The ability to detect paradoxes lies at the root of the scientists' propensity for puzzlement. An appreciation of the paradoxical leads the scientist to generate novel concepts and models which direct productive research efforts.

What is argued is that the researcher's skill in perceiving paradox is an essential factor underlying scientific progress. Such paradoxes often lead to competing perspectives or approaches with regard to the issues involved. Finkelman suggests that such a situation involving competing perspectives is peculiar to psychological research, and that change in psychological perspectives is often misclassified as progress. Finkelman contends that, contrary to prevalent claims, there has been little if any cumulative progress in psychology over the years while, in contrast, such progress does occur in the "hard core" sciences such as physics.

It is here contended, in opposition to Finkelman, that in order for progress to take place, it need not be the case that "newer views be refinements of the older ones" (2, p.185). Progress is here conceptualized in terms of advances in the formulation of questions, and not simply in terms of more refined answers to traditional, long-standing problems, (questions). To formulate more refined questions, it is often necessary to adopt a new perspective and hence the latter, it is held, should not be viewed as a regressive step. Such a process, it is suggested, is common to all scientific disciplines. The view expressed here was more eloquently advanced by Einstein in the following statement:

> The formulation of a problem is often more essential than its solution which may be merely a matter of mathematical or experimental skill. To raise new questions, new possibilities, to regard old problems from a new angle, requires creative imagination and marks real advance in science (3, p.92).

It is necessary to consider how it is that perceived paradoxes often lead to productive research addressing novel questions. Here follow then a few cases in point which illustrate the mechanisms involved in such a process.

CASES IN POINT

A Paradox in the Psychological Literature Pertaining to Depression

Abramson and Sackeim (4) discuss a paradox arising in the research literature on depression. There exists a theoretical paradox in that depressives are conceptualized as persons willing to blame themselves for events which were, or are believed to have been, beyond their personal control. That is, depressives are viewed as self-blaming individuals, guilt-ridden and ready to take personal responsibility for negative outcomes, while at the same time being fatalistic and tending to believe that negative outcomes are beyond their personal control.

According to Abramson and Sackeim, this theoretical paradox is not simply due to "semantic confusion, and mislabelling or the falsity of one or more of the theories" (5, p.843). Rather, these authors hold that since this conceptual paradox is also reflected on an empirical level, it is a valid one. That is, depressives do tend, in a variety of experimental contexts, to espouse beliefs in the uncontrollability of outcomes while also assuming personal responsibility for those outcomes, especially if they are negative. In addition, depressives often engage in self-punitive behavior for the same (6, 7, 8).

Abramson and Sackeim suggest that the tendency to assume responsibility for events presumed to be uncontrollable may be a "specific feature of depression or a general phenomenon, highlighted in depression" (9, p.845). Several research studies appear to indicate that there is a tendency to assign blame to persons on bases other than their actual degree of potential control over negative outcomes and their willingness to take advantage of such control in order to avoid these negative outcomes. For instance, Walster (10) found that attribution of responsibility for an accident was *not* a function of the facts in the case relevant to the making of judgments regarding potential control. Rather, attributed responsibility was found to vary in accordance with the severity of the negative consequences of the accident. Less responsibility was assigned when the consequences of the accident were trivial compared to the case in which they were severe. A study by Jones and Aronson (11) revealed similar findings. The Jones and Aronson study dealt with attribution of fault to a rape victim. This study revealed that attribution of fault to a rape victim was greater the more respectable the victim was, for instance a married woman as opposed to a divorcee. Such studies indicate a general tendency or readiness to assign responsibility for negative outcomes based upon factors which bear little or no relation to the controllability of those outcomes.

The paradox in the literature on depression is a fruitful one to consider in that it further elucidates the complexity of the mechanisms underlying depression. Attempts to resolve the paradox have led to research examining whether beliefs in controllability and self-blame are primary symptoms or secondary functional manifestations of the disorder (12). The paradox in the depression literature may lead to attempted resolutions which serve to integrate diverse areas of psychological research. There are, for instance, possible links between the depressive's paradoxical orientation toward responsibility and the way in which people generally attribute responsibility to others for negative outcomes: "In general, in making judgements of the responsibility of others, the significant determining factor . . . is not whether the outcomes are controllable or uncontrollable" (13, p.849). It appears that there is a general tendency to make attributions of responsibility in such a way as to maintain the "illusion of control" (14). The notion of "illusion of control" is reminiscent, as

Abramson and Sackeim point out, of Freud's (15) contention that the infantile belief in omnipotence often influences the behavior of adults — be they diagnosed as psychopathological or not. Regardless of the final fate and interpretation of the paradox in the depression literature, what is clear is that the puzzle has stimulated much productive research and helped to delineate novel questions. In addition, the need for clarification of such seemingly straightforward notions as "controllability" has become apparent.

A Paradox in the Literature on Perceived Freedom: Illusions of Freedom Created by Means of Behavioral Control Techniques

According to Skinner (16), it is possible to create an "illusion of freedom" by means of positive reinforcement in that under positive reward conditions, the individual often fails to recognize the controlling effects of the reinforcer. A paradox exists then in that an individual under the influence of powerful controlling factors subjectively experiences himself to be free. Skinner attempts to resolve this paradox by attributing the phenomenon to man's proclivity for irrational modes of thought. Such an illusion may, however, play a functional role.

The evidence would seem to suggest that the "illusion of freedom" when it does occur — and there is likely to be controversy on this point — is a most rational illusion, however much such a suggestion may appear to be a contradiction in terms. It is a rational illusion insofar as it leads to effective coping behaviors in circumstances in which nonillusory views of the world may be quite detrimental to the individual's mental and physical well-being. For instance, if an individual were to find himself in a situation involving negative stressors over which he had no control, an illusory perception of control or of freedom of choice would result in a certain resistance to the stress. In support of such a view is the finding of Glass and Singer (17) that performance on proofreading and problem-solving tasks for subjects exposed to uncontrollable loud noise improved considerably if they *falsely* believed that a person who could turn off the noise was accessible. This was so even though the noise was not turned off. Such studies led Zimbardo to suggest that there is:

> a psychologically functional value for human superstition. Superstitions prevent learned helplessness by providing the superstitious person with an "illusion of control" (18, p.399).

A distinction should perhaps be made between what might be termed "cognitive control" as opposed to "environmental control". An individual may, insofar as the objective situation is concerned, have no control whatsoever over events, yet exercise a degree of cognitive control to the extent that he imagines himself to be a free agent. This cognitive control may, in particular instances, have beneficial effects such as modification of the typical physiological responses to stressful events. For instance, cognitive control may allow the individual to return rapidly to baseline heart rate levels at the termination of the negative stressor. This then would tend to reduce the risk of chronic anxiety, and the negative physiological response patterns which such a syndrome entails.

Cognitive control may also mediate changes in the response to pain. It is well known that the individual's attitude toward pain is a factor which influences its perceived intensity. For example, Zimbardo (19) demonstrated that individuals who chose, under little social pressure, to continue in a task where they would be exposed to painful shock perceived the shock to be of a lesser intensity than they had originally. Zimbardo suggests that in an attempt to rationalize their seemingly illogical decision to continue exposing themselves to pain, subjects via cognitive control over their perceptions, were able to experience the shock as less intense. Additional evidence appears, in Zimbardo's view, to suggest that these subjects responded less physiologically to the shock and were not simply verbally reporting less pain.

Studies by Fields may provide an explanation for how it is that cognitions mediate a reduction in pain. Fields demonstrated that subjects who were placebo reactors experienced a worsening of pain when administered naloxone, a pure narcotic antagonist. Naloxone did not cause a worsening of pain in subjects who do not respond to placebos. Fields concludes tentatively that: "Since naloxone is a pure narcotic antagonist . . . the conclusion from the study is that the placebo effect is based on the release of endorphins (endogenous morphine)" (20, p.172). Fields raises the question: "Assuming all people have endorphins in their brains, why do only one-third of the patients get relief from placebos?" (21, p.172). It may be that the answer to Field's question is that only certain individuals can appropriately mobilize a sense of cognitive control over pain which results in activation of the brain's own analgesia system. The precise mechanisms by which such hope for relief might be translated into activation of the analgesia system and release of endorphins is by no means clear. However, such findings provide evidence which suggests that perceived control, even if illusory, may sometimes be part of an adaptive response pattern. Field's findings further point out how difficult it is to determine at which point illusion begins and ends, since it now appears that placebo effects may have a biochemical basis, and that faith in the efficiency of placebos is not entirely misplaced.

Research efforts have been directed toward determining both the effects of perceived control (i.e. freedom of choice) and the range of situations in which such cognitive control appears to be activated. As Skinner (22) points out, a sense of freedom may be aroused when an individual is induced to behave in certain ways via positive reinforcement. However, further evidence suggests that persons are more likely to attribute freedom to themselves and others when behavior is consistent with presumed predispositions (23). The ability of positive reinforcers to induce a sense of freedom may thus appear to be somewhat limited.

Several studies have demonstrated high positive correlations between perceived choice and feelings of competence and enjoyment (24, 25, 26). Relationships have also been found between perceived control over behaviour in a future situation and perceived choice in the making of the decision which gave rise to that situation (27). The mechanisms involved in perceptions of freedom are open to debate. What is clear is that the paradox represented by the illusion of freedom has given rise to a fruitful area of research directed toward discovering the determinants and consequences of perceived freedom. Future research may be directed toward delineating, for example:

> the conditions under which people try to present themselves to others as either free or not free . . . how environmental conditions such as crowding and lack of privacy influence people's sense of freedom . . . the role of perceived freedom in therapeutic and educational endeavours (28, p.94).

The Paradoxical Coleman Report: Absence of any Effect Due to Quality of the Educational Environment

Numerous studies appear to indicate that improvements in the quality of education do not result in any measurable changes in achievement (29). McClelland comments upon such findings as follows:

> All sorts of strange inferences have been drawn from this intuitively absurd conclusion, not the least of which is a feeling that there is no use in trying to improve education or spend more money on it, since the quality of education doesn't make a difference. Teachers know better. Parents know better. Pupils know better, . . . The problem has been that educational measurement has been dominated by a technology in which a student demonstrates what he knows by choosing among a variety of preset alternatives to a question . . . A more appropriate technology would seem to require measures of how a student thinks or what he chooses spontaneously to do after he has been exposed to quality education, because it is these types of behaviour that such education aims to influence and that matter in life (30, p.207).

The apparent paradox which arises from the findings cited in the Coleman report led to a reexamination of the assumptions upon which such research is based. One of the report's most incisive critics is Ryan who describes it as "a triumph of sophisticated research design over common sense" (31, p.43). According to Ryan, what Coleman did was to deal with different "racial" or ethnic groups separately on the grounds that analyzing them together might "cause any school characteristics highly associated with race or ethnicity to show a spurious relation to achievement" (32, p.311). Also controlled for was family background. Coleman's report concluded that "differences in school facilities and curriculum . . . are little related to differences in achievement level of students" (33, p.316).

Coleman thus compared school expenditures among schools in white upper middle class areas to see if the different expenditure levels were reflected in differential achievement outcomes. A separate comparison was made for expenditures in poor black neighbourhood schools. It is, according to Ryan, not surprising then that Coleman found that expenditures were of little import. Having accepted the validity of the aforementioned conclusion, Coleman went on to attribute the differential in achievement levels between black and white students to "the background culture from which these groups come". This despite the evidence, for example, of high educational aspirations among black students (34, p.280).

The apparent paradox which emerged from the Coleman report and others similar to it revealed, in the view of the report's critics, certain of the biases operating in much of the educational research addressed to highly political issues such as the question of the effects of integration upon school achievement. In any case, the analysis of Coleman's paradoxical findings led to a renewal of interest in the effects upon student achievement of the expectations for success or failure communicated by the system, teacher, administrators and others, and internalized by the student. Much valuable work has been done since the Coleman report on the effects of classroom and school atmosphere upon student performance.

An ambitious and innovative project on the effects of classroom emotional climate upon achievement is that conducted by De Charms (35). De Charms aimed to train teachers to discriminate between teaching styles that encourage student involvement in classroom learning and those that don't. A longitudinal study was undertaken in which:

> teachers were encouraged to understand and experience themselves as causal agents . . . The teachers then worked with the project staff to develop units for classroom use emphasizing the awareness of self-concept, achievement motivation, realistic goal-setting . . . The goal of these units was to increase students' motivation by helping them to experience themselves as the causal locus of their own behaviour (36, p.34).

De Charms' approach was a systems one in which both teacher and student were held responsible for learning outcomes. The need for restructuring of the psychological environment in the classroom was emphasized as a prerequisite to any desired changes in the students' conceptions of school learning and its personal significance or lack thereof.

> In other words, students were encouraged to become self-invested in classroom learning through situational restructuring that permitted them to experience having some control (37, p.34).

The findings of the training project revealed that students of trained teachers were less often late or absent, exhibited greater gains on standardized achievement tests, and higher Origin scores on a measure of the Origin-Pawn variable developed by Plimpton. In addition, student responses discriminated between trained and untrained teachers on a classroom climate instrument indicating that trained teachers responded to students differently (38).

Additional studies demonstrated that simply equipping teachers with the knowledge as to those teaching styles that inhibit or promote student investment in classroom learning is not always enough to alter teaching styles where inappropriate. These findings led De Charms to conclude: "that future training programs . . . must encourage participants to move beyond cognitive understanding to want, to know how, and to try to invest in personal change" (39, p.38).

The emphasis has since the Coleman report, for the most part, shifted away from "blaming the victim" to examining the situational factors underlying poor academic performance and how these interact with individual personality variables (40). Aside from its leading to an analysis of some of the critical factors involved in the problem of performance differentials between various identifiable groups of students, this shift may ultimately foster certain changes in the student's self-perception. A change in official approaches to the problem may permit the low achieving student to recognize the situational factors responsible in part for his difficulties.

A situational attribution, even if but partial, is likely to result in a reduction in anxiety and self-deprecating behavior. Support for the latter contention comes from a variety of studies drawn from the attribution literature (41). According to Storms and McCaul's model "the attribution of undesirable behaviors to dispositional qualities of the self creates anxiety and this anxiety subsequently increases the occurrence of the negative behavior" (42, p.155). For example, using the emotional exacerbation model as a base, these authors demonstrated that the rate of disfluencies in the speech of normal speakers could be markedly increased over baseline levels. This was accomplished as the result of individuals being led to attribute the disfluencies to the self, to regard them as a reflection of their own personal speech pattern and ability. Individuals led to make situational attributions, that is encouraged to attribute the speech disfluencies to the fact that they were in an experiment, a stressful situation, did not show such increases in stammering.

The exacerbation model may also be applicable to school achievement. To the extent that a student assumes total responsibility for failures, anxiety is increased and the student expects to be defeated in future. Such a view is quite consistent with the findings of Weiner and Sierard (43). These authors found that misattribution of failure to an external cause resulted in improved performance on a digit-symbol substitution task for low achievers, while such an attribution decreased performance for subjects high in need for achievement. The authors suggest that "changes in cognitive networks are sufficient to produce changes in achievement strivings" (44, p.149).

These results would seem to illustrate the value of De Charms' approach-enhancing achievement motivation via changes in causal ascription. It appears that the lower achiever is one who experiences great shame in failure, and little pride in success (45, p.85). By means of attribution training perhaps both the student as well as the teacher can come to more realistically evaluate the extent and nature of their responsibility for particular learning outcomes. Precisely which types of attribution will be optimal for particular individuals at given times in their history is a complicated issue as evidenced by the Weiner and Sierard research.

The paradox highlighted by the Coleman report has led to a reappraisal of what are the critical factors underlying school achievement. However, the controversy is still alive and well, and far from settled, as evidenced by current debates concerning high school "competency testing" in the United States. Once again, it is not clear whose competencies ought to be tested — the teachers' or the students'. The answer depends, in large part, upon to whom is attributed the responsibility for the low academic achievement of students.

On the issue of what determines school achievement levels, Bloom contends that: "There is at present little support for believing that the characteristics of teachers, classrooms, or schools have much effect on the learning of students". . .Bloom goes on to qualify his statement by suggesting that it is "the teaching and not the teacher that is central, and it is the environment for learning in the classroom rather than the physical characteristics of the class and classroom that is important for school learning" (46, p.111). Whether or not Bloom's view is but a restatement of Coleman's thesis, or represents a subtle but significant shift in conception, is a debatable and hopefully empirical question.

Bloom's emphasis upon the quality of instruction as a major determining factor of achievement levels would seem to suggest that he is relatively optimistic regarding the possibility of improving student-teacher interactions. For Bloom, quality instruction provides appropriate cues to the learner, allows for learner participation and involves adequate reinforcement and feedback. In his view, the quality of instruction accounts for at least one-fourth of the variance on relevant cognitive achievement measures (47, p.108). His position has some relation to that of Ryan insofar as Bloom too contends that "Who can learn in the schools is determined to a large extent by the conditions in the school; the quality of instruction is a major determiner of who will learn well — the few or the many" (48, p.138). Bloom's research has centred around providing students with the necessary cognitive and affective prerequisites for learning, as well as the modification of instruction strategies. When such factors are equalized for students there is, according to Bloom's research, a considerable reduction in variability in achievement levels. He contends that the implication of his theory is that the effects of "school and the learning process are capable of being very strong — so strong that individual learners can be made either equal or very unequal with regard to particular learning outcomes" (49, p.213).

While Bloom contends that reports such as the Coleman report indicate that "the reality of school effects has been far from its potential" (50, p.213), Ryan would no doubt adamantly object that school effects have already been operative and account for most of the inequities which do exist between students. Whatever one's bias, it is evident that the research, stimulated in part as a reaction to the Coleman Report, has served to reopen the question as to what degree the social structure and emotional climate of the school is responsible for students' learning outcomes. Hence, the paradox formulated in the Coleman report, whether but apparent or not, has served to instigate complex, controversial and needed lines of educational research.

A Final Note on the Value of Paradoxes

What is evident is that paradoxes, whether well-founded or simply misleading, serve important functions in the research process. They often lead to the disclosure of hidden assumptions, imprecise concepts, and unchallenged but tenuous views. Most significant perhaps is the fact that paradoxes often reveal the contradictions embedded in what we naively term "solutions". They serve to illustrate that all empirical solutions are in the final analysis but apparent; regardless of how cognitively satisfying they may be. In weakening traditional solutions, paradoxes open up the possibility for new research directions and redefinitions of the basic questions to be addressed.

REFERENCES

1. Polanyi, M. and Prosch, H. *Meaning*. Chicago: University of Chicago Press, 1975.
2. Finkelman, D. Science and Psychology. *American Journal of Psychology*. 1978, Vol. 91, No. 2, p. 179-199.
3. Einstein, A. and Infeld, L. *The Evolution of Physics*. New York: Simon and Schuster, 1938.
4. Abramson, L.Y. and Sackeim, H.A. A Paradox in Depression: Uncontrollability and Self-Blame. *Psychological Bulletin*, 1977, Vol. 84, No. 57, p. 838-851.
5. Ibid., p. 843.
6. Hiroto, D.S. and Seligman, M.E.P. Generality of Learned Helplessness in Man. *Journal of Personality and Social Psychology*, 1975, 31, p. 311-327.
7. Beck, A.T. *Depression: Clinical, experimental, and theoretical aspects*. New York: Harper and Row, 1967.
8. Forrest, M.S. and Hokanson, J.E. Depression and autonomic arousal reduction accompanying self-punitive behaviour. *Journal of Abnormal Psychology*, 1975, 84, p. 346-357.
9. Abramson, and Sackeim, *A Paradox in Depression*, p. 83.
10. Walster E. Assignment of Responsibility for an Accident. *Journal of Personality and Social Psychology*, 1966, 3, p. 73-79.
11. Jones, C. and Aronson, E. Attribution of Fault to a Rape Victim as a Function of Respectability of the Victim. *Journal of Personality and Social Psychology*, 1973, Vol. 26, No. 3, p. 415-419.

12. Abramson and Sackheim, *A Paradox in Depression,* p. 838-851.

13. Ibid., p. 849.

14. Kelly, H.A. Attribution in Social Interaction. In Jones, E.E., Kanouse, D.E., Kelly, H.A., Nisbett, S., Valins, S., and Weiner, B. (eds). *Attribution: Perceiving the Causes of Behavior.* New York: General Learning Press, 1971, p. 1-26.

15. Freud, S. *Civilization and its Discontents.* Strachey, J. (ed. and trans.), New York: Norton, 1961. (Originally published 1930).

16. Skinner, B.F. *Beyond Freedom and Dignity.* New York: Knopf, 1971.

17. Glass, B.C. and Singer, J.E. *Urban Stress: Experiments on Noise and Social Stressors.* New York: Academic Press, 1972.

18. Zimbardo, P.G. and Ruch, F.L. *Psychology and Life.* Diamond Printing, Ninth Edition, Gleview: Scott Foresman and Company, 1977.

19. Zimbardo, P.G.; Cohen, A.; Weisenberg, M.; Dworkin, L.; & Firestone, I.; The Control of Experimental Pain. In Zimbardo, P.G. *The Cognitive Control of Motivation.* Glenview: Scott, Foresman and Company, 1969. p. 100-125.

20. Fields, H.L. Secrets of the Placebo. *Psychology Today,* Nov. 1978, p. 172.

21. Ibid.

22. Skinner, B.F. *Beyond Freedom and Dignity.*

23. Kruglanski, A.W. and Cohen, M. Attributed Freedom and Personal Causation. *Journal of Personality and Social Psychology,* 1973, 26, p. 245-250.

24. Jellison, J.M. and Harvey, J.H. Perceived freedom in every day decisions. Unpublished manuscript, University of Southern California, 1975.

25. Harvey, J.H. and Jellison, J.M. Determinants of perceived choice, number of options, and perceived time in making a selection. *Memory and Cognition,* 1974, 2, p. 539-544.

26. Brock, T.C. and Becker, L.A. Volition and attraction in every day life. *Journal of Social Psychology,* 1967, 72, p. 89-97.

27. Harvey, J.H. and Harris, B. Determinants of perceived choice and the relationship between perceived choice and expectancy about feelings of internal control. *Journal of Personality and Social Psychology,* 1975, 31, p. 101-106.

28. Harvey, J.H. Attribution of Freedom. In Harvey, J.H., Ickes, W.J. and Kidd, R.F. (eds.) *New Directions in Attribution Research,* Volume 1, New York: Lawrence Erlbaum Associates, 1976.

29. Coleman, J.S. *Equality of Educational Opportunity.* Washington, D.C.: U.S. Office of Education, 1966. (OE#38001).

30. McClelland, D.C. Managing Motivation to Expand Human Freedom. *American Psychologist,* March 1978, p. 201-210.

31. Ryan, W. *Blaming the Victim.* New York: Vintage Books, 1971.

32. Coleman, *Equality of Educational Opportunity.*

33. Ibid., p. 316.

34. Ibid., p. 280.

35. De Charms, R. *Enhancing Motivation: Change in the Classroom*. New York: Halsted Press, 1976.

36. Cohen, M.W. Emrich, A.M.; and De Charms, R.; Training Teachers to Enhance Personal Causation in Students. *Interchange,* 1976-77, Vol. 7, No. 1, p. 34-38.

37. Ibid., p. 34.

38. Ibid.

39. Ibid., p. 38.

40. Rutter, M. et. al. *Fifteen Thousand Hours: Secondary Schools and their Effects on Children,* London: Open Books, 1979.

41. Storms, M.D. and McCaul, K.D. Attribution Processes and Emotional Exacerbation of Dysfunctional Behavior. In Harvey, Ickes and Kidd, (eds.) *New Directions in Attribution Research,* p. 143-164.

42. Ibid., p. 155.

43. Weiner, B. and Sierard, J. Misattributions for Failure and the Enhancement of Achievement Strivings: A Preliminary Report. In Weiner, B. (ed.) *Achievement Motivation and Attribution Theory*. Morristown: General Learning Press, 1974, p. 140-150.

44. Ibid., p. 149.

45. Kukla, A. Attributional Determinants of Achievement-Related Behavior. In Weiner, (ed.) *Achievement Motivation and Attribution Theory,* p. 81-92.

46. Bloom, B. *Human Characteristics and School Learning*. New York: McGraw-Hill, 1976.

47. Ibid., p. 108.

48. Ibid., p. 138.

49. Ibid., p. 213.

50. Ibid.

NOTES

1. The Origin-pawn distinction is similar to the internal-external locus of control categories. De Charms defines these terms as follows: "To the extent that one experiences a causal influence upon the events of one's life, one is an Origin. Insofar as one feels causally separated from the events of one's life, one experiences oneself as a Pawn. A scale to measure this variable was developed by Plimpton, F.A. and reported in an unpublished manual, Washington University, 1969.

CHAPTER IV

THEORETICAL BLINDERS: SETTING ARTIFICIAL LIMITS

Psychologists, as do scientists in general, often tend to place artificial limits on the research process by prematurely designating particular topics and research strategies illegitimate. In this section are examined various examples of the operation of theoretical biases which have restricted the range of problems considered and perspectives held. The focus in the following discussion is upon the research which finally exposed these biases and to a degree served to weaken their influence. Also considered is the influence of the cultural context in which the scientist operates. The complexity of scientific data, and the myriad of possible interpretations which it permits is also to be underscored in what follows. It will become evident that observations do not directly attest, in any simple way, to the validity of a particular model.

RE-EMERGENCE OF THE CONCEPT OF CONSCIOUSNESS

It has been suggested by certain philosophers that the social science paradigm of persons, rather than contributing to a deeper understanding of ourselves, has resulted in

> an amassing of more and more details concerning the periphery of persons . . . a superficial, external point of view which ignores the experience of actual, particular, unique individual persons who are conscious of themselves as subjects of experience, not as mere external objects or passive things (1, p.206).

Contemporary psychologists - not all, but a goodly proportion — have addressed themselves to the question of self-consciousness; its origins and consequences. Such research would seem to weaken the claim of the philosopher Rabb, that "the very nature of scientific observation forces us to ignore the fact that we are beings endowed with self-consciousness or self-awareness" (2, p.206). Indeed, the intent of this book is to point out the insights scientific observations in the psychological laboratory provide into the nature of the scientist's cognitive processes. Scientific observation itself then, it is here suggested, can provide a framework within which the scientist may engage in self-reflection.

It is no longer the case that the problem of self-awareness is considered to be outside the purview of proper scientific questions. Such an assumption characterized the positivist conception of science but is no longer tenable — if indeed it ever was — given the fruitful work in this area. Here follows an examination of but some of the literature pertaining to the study of consciousness. Those studies have been selected which have particular implications for notions of what is a person.

The Notion of a Duality of Consciousness

Studies with individuals with intact brains have provided evidence which has been interpreted as indicating differential specialization of function between the hemispheres. For instance, in dichotic listening situations the right ear (left hemisphere) is generally superior with verbal material; the left ear (right hemisphere) is often superior with music (3). Such specialization of function is strikingly revealed in a study in which it was found that the direction in which a person gazes is affected by the kind of question asked. If the question is verbal-analytical (divide 36 by 2) there is more eye movement to the right than if the question involves spatial thinking (which way does the Indian face on a nickel) - (4). When a person is writing, there is more alpha rhythm in the right hemisphere than in the left while when arranging blocks more alpha is recorded from the left hemisphere than the right. Since alpha is generally considered to be a sign of restfulness, the finding has been interpreted as evidence that "we turn off the hemisphere not involved in the situation" (5, p.33). It is Ornstein's view that "in most ordinary activities we alternate between the two modes (of consciousness) selecting the appropriate one and inhibiting the other" (6, p.33). Ornstein conceptualizes the left hemisphere as being the more logical and analytical, the right the more intuitive and spontaneous.

While Ornstein speaks of alternative modes of consciousness within a single individual operating perhaps in parallel; perhaps alternatively, others have suggested that there exists two consciousnesses and hence two persons within each of us. The evidence generally used to support the latter view comes from the study of split-brain patients. These are patients who have had the major nerve fibre tract connecting the hemispheres, the corpus callosum, severed. Under laboratory conditions, such patients appear to display behaviors consistent with the notion that they are operating with two independent consciousnesses, each unaware of the other's existence. For instance, if a picture of a nude female is presented to such a split-brain patient in such a way that the information is relayed to the right hemisphere only, the patient responds by blushing. When asked what are the reasons for the apparent uneasiness, the patient is unable to offer a verbal explanation. Characteristically, such subjects report having seen nothing or just a flash of light: "Apparently only the emotional effect gets across (to the left hemisphere), as if the cognitive component of the process cannot be articulated through the brainstem" (7, p.65). Sperry concludes that such evidence supports the notion of two separate independent conscious entities for the most part unaware of one another. Two persons incorrectly presumed to be one (8, p.65). Sperry does not however extrapolate this notion to those humans with intact brains.

Such behavioral studies by clinical neurologists, psychophysiologists and others are directly relevant to the question of what are the appropriate criteria for personhood and for personal identity. Does each separate consciousness originating in each hemisphere qualify as a distinct person? If so, what becomes of the personal identity of the individual harbouring these independent conscious entities? Are we to think of the split-brain individual as a disintegrated personality or a unified single person capable of switching modes of consciousness? Perhaps the foregoing questions are themselves misleading and will be replaced by more fruitful ones. In any case, such questions have stimulated a lively controversy which has brought to light just how muddled some of our most basic concepts are, concepts such as "person" and "action". At this point, it is valuable to examine the various proposed criteria specifying what it means to be a person and to consider also the objections which have been raised to these.

Dewitt (9) argues that only the speaking hemisphere is self-aware and so qualifies as a person as is the case for cerebrally intact humans as well. Others, such as Puccetti (10), suggest instead that the split-brain studies provide evidence for two persons each self-aware as well as evidence for two minds. Puccetti supports his view by pointing out that the right hemisphere does have language skills - a feature Dewitt takes to be a prerequisite for self-awareness-though it is unable to translate these into speech or writing. For instance, the right hemisphere can guide the retrieval of related objects when given abstract verbal definitions such as "measuring instrument", "eating utensil" (11).

Whether the existence of language is the most appropriate indicator of self-awareness is a matter of continued controversy. There are several pieces of evidence which have been taken by some to indicate that the right hemisphere is in fact self-aware despite its limited linguistic capabilities. For instance, "in cases where the right hemisphere has had, say a key presented to it visually and the patient is asked what he has seen, the speech hemisphere often guesses while the head nods negatively at such wrong guesses, or frowns disapprovingly. What is this if not a conscious, wilful reaction on the part of the right hemisphere?" (12, p.342). Evidence for the self-awareness of the right hemisphere is alleged in the study conducted by Levy, Trevarthen and Sperry (13). In the latter study, different half faces were projected to the separate hemispheres. Before the subject could *point* to the face he had seen (right hemispheric function), the visual stimulus array was removed and the subject was asked to first *say* what he had seen (left hemispheric function). Or conversely, before the subject could say what he had seen, he was instructed to first point to the correct face. The verbal response thus conflicted with the manual response.

> When the verbal response came first under direction from the left hemisphere, the right hemisphere had to choose between the announced verbal selection and its own strong impression favoring one of the faces. In only one trial in four did this result in loss of the response preferred by the right hemisphere. When the manual response came first, the verbal hemisphere had to decide whether to be consistent and describe the face that had already been pointed to or ignore this in favor of its own recall of a different face . . . Omission of all reference to the right half face occurred only once in four trials (14, p.67-68).

The interesting aspect of these results is that each of the hemispheres demonstrates an ability to choose between competing alternatives that which best fulfills its goal - selecting the correct picture. Each hemisphere appears to operate upon its own knowledge base, ignoring for the most part the intrusions of the opposite hemisphere, and hence avoiding conflicts. Generally speaking, the split-brain individual is as surprised as is the experimenter when contradictory behaviors are engaged in; as when the patient shakes his head "no" in response to the nonverbal responses executed by the right hemisphere. The patient in such an instance appears to consider himself to be a single unified person and not the embodiment of two persons each with a mind of his own: "It seems to it (the left hemisphere) as if its body was inexplicably out of step with its intention" (15, p.349). The left hemisphere may in this way simply be expressing biases fashioned by traditional concepts of what is a person. Whether the right hemisphere can better accept the notion of two persons coexisting within a single body is an open question.

There are several factors which in fact serve to create the impression of a unified consciousness for the split-brain individual.

> The fact that these two separated mental spheres have only one body, so they always get dragged to the same places, meet the same people, and see and do the same things all the time and are thus bound to have a great overlap of common, almost identical experience (is one unifying factor) (16, p.723-33).

Whether the impression of a unified consciousness - be it in the split-brain individual or the subject with the intact brain - is an illusion or not is an intriguing and as yet unresolved issue. Eccles, for instance, suggests that the neurological evidence indicates that what self-awareness the right hemisphere does display results from the transmission of information via nonspecific pathways to the left, dominant hemisphere (17). This, however, still leaves open the question of how it is that the right hemisphere can react negatively to suggestions made by the speaking hemisphere, and abide by its own decision as in the study by Levy, Trevarthen and Sperry mentioned earlier.

Indeed, even if the right hemisphere were not self-aware, it is not clear whether it would not yet qualify as a person. There are innumerable instances in which the individual with an intact brain exhibits a lack of self-awareness, but few of us would be willing to challenge his status as a person as a consequence. For instance, persons with intact brains are generally unaware of the cognitive processes involved in memory retrieval. The notion of awareness is itself problematic for there appear to be levels of awareness and which level one must attain to reach personhood is unclear. Consider the example of the operation of such levels of awareness provided by Polanyi. Polanyi uses the term "nonfocal attention" to describe our level of awareness of the two-dimensional pictures we integrate to create a three-dimensional stereoscopic image (18, p.29). Does such "nonfocal attention" involve self-awareness nonetheless? While we are aware of ourselves viewing the three-dimensional stereo picture, we do not in any simple sense appear to be aware of ourselves scanning the separate two-dimensional images nor of computing their merger. Persons with intact brains then often execute behaviors of which they are not fully aware, just as the split-brain individual is not always fully aware of what he's up to (i.e. the left hemisphere is unaware generally of the reasons for particular right hemispheric responses).

According to Robinson (19), the split-brain data are simply another instance of parallel processing, which is by no means restricted to split-brain patients. Robinson cites Sperling's (20) studies on the span of visual memory as evidence for such parallel processing in the cerebrally intact human. In Sperling's studies, subjects could report from memory between five and nine of sixteen items briefly presented visually in a matrix. However, when a cueing ring was projected over one of the matrix locations immediately after the display disappeared, the subject was able to report the letter previously occupying that position even though on the trial in which the letter was presented he failed to report it.

> The observer simultaneously 'knows' and does not 'know' that the letter in question was part of the array. Without the array, he will swear he has reported everything he has seen. With the cue his span of retention is widened, ex post facto. If trials without cues correspond to our observer in state-A and those with cues to him in state-B, we must conclude that he does not know in A what he reliably discloses in B. On Puccetti's account two minds, and therefore two persons are operating (21, p.74).

Robinson prefers the interpretation that in Sperling's studies, as in the case of split-brain subjects, persons are not verbally reporting all they know or have perceived.

The right hemisphere may be self-aware as evidenced by its ability to override suggestions from the left hemisphere, for example. However, several theorists, as has been discussed, argue that this need not imply that there are two persons within a single body. Note that there would appear to be but one motive exhibited by both hemispheres, and that is to comply with the task demands and the experimenter's instructions. It may be simply the case that each hemisphere in the split-brain patient has a separate memory store, and that there exists no access to the opposite hemisphere's store. Therefore, each hemisphere may make different judgments based on differing information. While each hemisphere may execute different *behaviors* to accomplish its goal (i.e. point to one picture as that previously presented while verbally signifying that the target picture was one other than that pointed to), the *action* involved may not be contradictory. That action, as mentioned previously, is that of complying with the task demands - attempting to select the correct picture. Hence, while each hemisphere may be responsible for apparently contradictory behavior, according to this interpretation, a single person is responsible for the act of attempting to comply with the experimenter's demands. "The cause of an act is quite simply the agent that performs it" (22, p.56). That agent, would be the integrated individual whose goal it is to assist the experimenter. There are then innumerable plausible interpretations of the findings. The split-brain data dictate not only a re-examination of the concept of "person", but also of the philosophical distinction between the concept "behavior" and "action", and its potential utility for clarifying the split-brain data.

Case reports of a patient who tied the belt of his robe with the assistance of the left hand together with the right, then promptly untied it with the left, may represent an instance of contradictory "actions" representing the conflicting intents of two agents. However in several such cases there was extensive damage of the right hemispheres (23). It is necessary then to examine how typical hemispheric conflicts are in split-brain patients in order to determine if such a patient can be described as one or perhaps two persons. In order to do so will first require an accurate analysis of what constitutes an "action", of which intentions are overriding, and the specification of the purpose of the action. Only then will it be possible to determine if in fact the patient is truly engaging in contradictory actions which embody competing and mutually exclusive intents. It is suggested here then that Sperry's conclusion that such cases provide evidence for "two free wills packed together in the same cranial vault" (24, p.304) is a premature one.

One of the reasons why consciousness regained its acceptability as a research area in the psychological community was the advent of computer modelling. The computer, it was and is still largely contended, could render notions such as memory and pattern recognition viable, precise, and meaningful. Finkelman has commented in this regard that:

> For many, the new arbiter appears to be the computer. When the computer was introduced, it became acceptable to talk about central processes. After subroutines were made available, it was all right to talk about mental analogies to *them.* It sometimes appears that the rate at which mental processes become legitimate objects of study is tied to the rate of progress in computer technology (25, p.190).

In the next section are examined some of the ways in which computer programmes have been applied in order to further understanding of human information processing. Controversies concerning whether computer modelling of human cognitive processes serves to broaden or limit psychological perspectives are discussed.

COMPUTER MODELS OF MIND

Neisser, in his classic text "Cognitive Psychology" states: "The task of a psychologist trying to understand human cognition is analagous to that of a man trying to discover how a computer has been programmed" (26, p.6). The fact that computer simulations are but analogies for the human cognitive processes modelled is a point, critics claim, which is often lost sight of. According to these critics, too often human cognitive processes are simply regarded as the running off of highly complex routines and subroutines, perhaps with possibilities for self-correction and some limited types of creativity. Such a conceptualization amounts, in their view, to nothing less than the reification of the computer programme. Sperry would, however, disagree. He goes so far as to suggest that programs can probably be devised which simulate human consciousness for in his view: "nothing excludes the possibility that consciousness be present in a machine or electronic device" (27, p.307).

The goal in the discussion which follows is to examine the nature of certain of the claims made regarding computer modelling of human cognitive processes; that is to consider the implications and underlying assumptions of such claims. Consider, for example, Sperry's contention that consciousness can very possibly be programmed into the computer or some such electronic device. Certainly, it is the case that computer programmes have already been devised which produce an *output* hardly distinguishable from that which would constitute the verbal responses of a conscious human being. Colby and his associates have, for example, devised programmes which simulate the verbal responses of neurotic individuals (28, 29). Yet, in what respects such a program is in fact a simulation of internal states of human consciousness is debatable.

The computer's states of consciousness, if such could be devised, would at some point be directly accessible to the programmer. In this respect, the computer's consciousness would be a faulty simulation of human consciousness, the latter being private and not fully communicable. One person may come to know indirectly something of the states of consciousness of another, as when a psychologist monitors the brain patterns accompanying different stages of sleep of one of his human subjects, or a husband experiences sympathy pains in harmony with his pregnant wife's labour pains. Such knowledge remains, however, indirect and based upon the observable correlates of internal states of consciousness. Since persons can have but indirect knowledge of the states of consciousness of other persons, at least at present, there is never any question as to shared consciousness. For instance, the husband experiencing sympathy pains is quite cognizant of the fact that the pain he is feeling is his own, not that of his wife, and that it is only his own pain of which he is directly aware, and again not that of his wife.

Suppose the computer programmer were able to devise a programme which resulted in a conscious machine. The emerging consciousness could not be said to belong uniquely to the computer given the fact that the programmer had direct access to such machine states of consciousness. Would there then be a *simulation* of human consciousness, or simply a shared consciousness distributed between programmer and machine? Perhaps in order that computers be developed with states of consciousness, the computer must be devised so as to experience its input, and not simply process it. In other words, such a development may require that the input have some emotional significance to the device which interprets it.

Some would argue that simulation attempts do not offer much insight into the human mind since "we need to know how human cognitive processes . . . work before we can build a machine of equivalent capability . . ." (30, p.227). That poses quite a difficulty in that there are many such processes of which we are unaware and cannot even at present formulate in a fashion which would make them amenable to empirical investigation. Of course, the opposite view is also often advanced that "by trying to program a machine with (cognitive) capacities we will perhaps improve our understanding" of those cognitive capacities (31, p.92).

One of the fundamental difficulties that researchers in artificial intelligence have yet to fully overcome is the problem of how to program in common sense. It has been found, for example, that

> mapping a two-dimensional pattern of light received from the environment onto structures that directly represent what is out there in the world can only be achieved if the machine is given some knowledge of the kind of objects that may be present . . . progress has been made in deriving accurate descriptions of three dimensional objects when some restrictions are placed on the type of object which may be present (32, p.260).

In other words, the computer programme must, if it is to be an efficient recognition device, contain common sense elements which specify, for instance, what types of objects it is likely to encounter in a particular environment.

Consider the role of common sense in chess playing and the difficulties involved in embodying the same principles in a programme for chess playing.

> We have it on Claude E. Shannon's authority that the game of chess can be completely formalized. Unfortunately, a computer playing a completely formal game, with the ability to consider 1,000,000 moves per second, would require 10^{95} years to make its first move. In other words, chess is formalized by making every move an instruction according to a rule, and in a forty-move chess game there would be 10^{120} such possible instructions. Since humans can play good chess and can plan long sequences of moves, it seems clear that they do not play formal chess . . . Hence, . . . the distinction between the informal process of learning to play a game and the formal process of learning its rules is valid (33, p.49).

Since Taube's words on the matter of computer programs designed to play chess were uttered, there has been considerable advance in the development of such programs. "Nevertheless . . . though both computer and brain sciences have made impressive strides, chess crowns still fit human heads" (34, p.12). The major stumbling block appears to be that it is still not known how it is that the human brain codifies and applies common sense hunches, "nor is it known whether such heuristics, or intuition, can be reduced to a step-by-step procedure" (35, p.15).

Puccetti (36) offers an interesting analysis of what might be occurring cognitively when a human plays chess, and is taking short-cuts based upon hunches. He begins his analysis by drawing a distinction between identification and recognition. By identification, Puccetti means an inference as to what is perceived based upon some kind of feature analysis. Such a process he takes to be the specialty of the left hemisphere which can execute verbal analytical descriptions of the input. Recognition, according to Puccetti, does not involve such inferential activity. It is a more efficient, rapid, and wholistic process carried out by the right hemisphere (37). Experiments by Brand (38), Ingling (39) and Posner (40) seem to illustrate the distinction between recognition and identification Puccetti is trying to make. These researchers found, for instance, that subjects could categorize an input as being either a letter or a number without being able to identify which letter or number it was. Similarly, it has been demonstrated that subjects can distinguish between words and nonwords very rapidly before having identified the word in question. Turvey concludes: "we should have to argue . . . that determining category membership is not based upon any simple or obvious feature analysis" (41, p.174). There may then be an experimental basis for the distinction Puccetti makes between identification and recognition processes. Recognition processes would be those rapid analyses such as are involved in assigning input into a letter or number category, processes which do not involve a detailed feature analysis. The output of such a process need not necessarily be verbalizable but may constitute rather a type of tacit knowledge.

In Puccetti's view, it is just such a nonanalytical recognition process which is critical in chess playing. He states:

> Humans . . . do not reason out (chess) problems at all. What they do is see the flow of pieces over the board in a dynamic way, leading to pressures and exchange, gains and losses of . . . material or security that will ultimately spell victory or defeat. A computer chess program cannot do this: it must substitute for the perceptual process an extensive search procedure . . . which may or may not allow it to achieve a similar game result (42, p.149).

It is this ability to organize the visual input appropriately which Puccetti holds is at the core of the advanced chess player's skill rather than "general knowledge of chess possibilities and probabilities" (43, p.150). The question arises then as to whether a program can be devised which captures such a recognition process. Puccetti contends that such is not a feasible project since programs "can only work with what we give them as visual decoding procedures, and we can only give them information-processing strategies modelled on the verbal hemisphere's quite inadequate ones" (44, p.145).

Thomas is considerably more optimistic about the prospect of devising wholistic type computer recognition programmes. He contends that:

> Computers as such are not committed to some particular form of reasoning — only the people who write programs for them are. There is certainly no reason in principle (or even in practice) why the Gestalt-synthetic point of view should not be embedded in a program (45, p.231).

Puccetti would no doubt counter that though it may be possible to program the computer with a set of step by step instructions for Gestalt type recognition, the *process* would not be a veridical simulation of Gestalt type recognition in humans. Humans, Puccetti argues, do not appear to recognize patterns by following such a set of formal rules. It is not clear how computers could be programmed other than by a set of such formal instructions. Whether Gestalt points of view can be programmed, however, remains an open question.

There is already in existence a robot which has impressive Gestalt-like perceptual abilities (46). This robot can construct visual models of objects in its environment, navigate through a set of obstacles in a room, and solve a number of problems requiring form recognition. What is not at all clear is whether such a robot simulates features of human pattern recognition. It may be that humans most often recognize objects first and then detect their features, rather than constructing object forms by integrating features according to a set of rules as does the robot described above. For instance, in recognizing a word persons generally recognize the word before the individual letters composing it, when the latter are presented successively (47). The perceptibility of the features (subcomponents) of a structure are generally enhanced if a meaning has already been assigned to the entire structure. For instance, letters are easier to discriminate when part of a word than when presented alone, a phenomenon termed the "word superiority effect" (48). It may be that persons formulate imaginative constructions of what might be present, and that such images guide feature analysis.

A relevant simulation of human perceptual processes, as any valid theory thereof, should include modelling of human perceptual illusions. It may be that equipped with the same knowledge base, the computer pattern recognition programme does *not* perceive illusions such as the Ames room size distortions. The human perceiver, who is told that the Ames room is a distorted trapezoidal room, rather than a rectangular one, is generally still subject to illusions of size distortion. Objects and persons located in the far corner of the room yet appear smaller than they ought be (49). The problem is not simply one of lack of relevant knowledge. Rather the perceiver's conscious experience is one of a rectangular room despite knowledge to the contrary. It is interesting to note that there is a resistance for persons to see their spouse distorted in size while their mate is walking in an Ames room (50). There are then emotional factors which determine in part the perceptual experiences of someone viewing objects in an Ames room.

It may be inappropriate to reduce the question of consciousness to the problem of intelligence. To equip the computer fully with human knowledge systems need not necessarily result in simulation of human conscious experience. As the example of the Ames room illustrates, conscious experience is often at odds with the knowledge base. While the computer programme, as the "human cognitive programme", may be able to acknowledge the fact that the room is trapezoidal, the latter cannot yet easily modify its perceptions accordingly. It may be then that human "mistakes" in, for instance, pattern perception most clearly reveal the qualitatively distinct nature of human consciousness. A most intriguing finding would be that of a computer *spontaneously* making similar errors — an indicator perhaps of machine consciousness.

Turing's test for deciding if a machine is intelligent involved placing the machine in one room and a person in the other. A questioner, via teletypewriter, communicates with each, not knowing the location of either. If the interrogator cannot reliably decide which is machine, which human then the machine, in Turing's view, is to be considered intelligent (51). While it is true that "we usually infer consciousness from observing intelligent actions" (52, p.18), such would appear not to be a foolproof strategy. Humans sometimes respond so as to contradict input information, as when the individual yet perceives a perceptual illusion despite knowing how it is created. Can computers be devised that produce output which violates the rules, knowledge base and common sense hunches which have been programmed in? Are not computer outputs more or less directly a function of the memory store or knowledge base so that they never contradict what is stored? Sloman would disagree. He states: "It is a myth that programs do just what the programmer intended them to do" (53, p.16). Computers engage "in behaviour that nobody planned and nobody can understand . . . such behavior leads to new possibilities being discovered" (54, p.16).

To return to the Ames room example, both machine and person might be able to provide the same intelligent answers regarding the fact that the room is actually trapezoidal and as such quite atypical. While both machine and person may have equivalent knowledge bases and memory stores, will only the person find it difficult to overcome the size distortion illusion created in the Ames room, despite possessing the relevant information which should negate such distortions? The person is subject to such illusions for they are an aspect of human conscious experience not directly related to the information at hand. In order that the computer "have" such an illusion may require then that the programme be devised such that it violates its own rules, creating an output which is illogical and in contradiction to its prior experience. This possibility may be open to empirical investigation.

Perhaps consciousness cannot be directly inferred from the apparently intelligent output of a computer (or of a person for that matter) for consciousness may not simply be an emergent property resulting from a knowledge system. "Much current work in A.I. is exploration of the *knowledge* underlying competence in a variety of specialized domains" (55, p.19). These programmes represent, in Sloman's words, "working theories" (56, p.16) of cognitive processes. It may be, however, that even if the knowledge embodied in such programmes were an accurate simulation of the major features of human knowledge systems, they would yet not generate consciousness. Future developments in artificial intelligence may provide some tentative answers to these questions.

It is not essential that a useful simulation model every feature of its referent, indeed "a simulation of a system that includes everything in that system is virtually useless as an aid to understanding the system: in the extreme case it is the system itself" (57, p.267). Yet it is necessary, it would seem, for the computer programme to contain an element of consciousness in order that it be justly held that "a particular programme really does explain some human ability, as opposed to merely mimicking it" (58, p.19). Unless, of course, one wished to deny the existence of consciousness as separate from behaviour as did Gilbert Ryle (59). In that case, one could properly contend that "the simulation of human brains by machines can be interpreted as the simulation of a machine by a machine" (60, p.76).

Sloman argues that "Insofar as anything clear and precise can be said about 'the way' in which a human being does something (e.g. plays chess, interprets a poem, or solves a problem) the appropriate procedure can in principle be built into a suitable simulation, so that we ensure that the machine does it in the same way" (61, p.110). It remains an open question, however, as to whether consciousness need be programmed in order that the simulation be a valid representation of an aspect of human cognitive processes or in fact if such can be accomplished. Sloman contends that "attempting pronouncements (about the limits of what can be done by computer programs) is about as silly as attempting to use an analysis of the printing process to delimit the kind of theories that will be expounded in textbooks of physics in a hundred years time" (62, p.105).

It would seem, however, that a discussion of possibilities and limitations is essential if the implications of what has been accomplished in A.I. are to be fully appreciated, and useful strategies for future goals are to be set out. There would appear to be a definite need for clarification given that A.I. researchers, such as Sloman, contend that computers are simply "artifacts to be improved and used" (63, p.106), while at the same time others hold that it may be possible to program computers with consciousness, a feature which would certainly transform the computer into a somewhat peculiar "artifact" indeed.

As the foregoing discussion testifies, cognitive processes are receiving much attention from researchers in the psychological sciences and related fields such as artificial intelligence. The "infiltration" of the cognitive view has taken place not only in the research domain but also in applied areas such as psychotherapy. Mahoney states that while it is seldom acknowledged in print, the proponents of each perspective may have come to recognize some of their ideological and practical limitations" (64, p.6). The development of "cognitive behavior modification" as a psychotherapeutic strategy, discussed next, represents an interesting example of how theoretical biases operate and may finally be challenged.

THE MODIFICATION OF BEHAVIORISTIC PERSPECTIVES IN PSYCHOTHERAPY

Several excellent texts and reviews are available (65, 66, 67) which discuss the encroachment of the cognitive view upon behavioristic therapeutic strategies. The determinants of this shift in view are partially revealed if one examines the technique of systematic desensitization. The traditional behavioristic format for systematic desensitization involved the client in imagining a hierarchically ordered, increasingly anxiety-provoking series of scenes. The client then would be trained to keep himself relaxed at each stage, and would gradually move up the hierarchy until ready to deal with the real life situation creating the difficulty. According to Wolpe, the originator of this technique, the responses of fear (anxiety) and relaxation are incompatible, and as a result the relaxation serves to inhibit the fearful response (68). What would seem apparent is that a cognitive element is involved in the technique insofar as imagery plays a key role. Often the therapist must train the client in the production of vivid images which are capable of generating the appropriate level of anxiety (69).

Locke, in analyzing Wolpe's psychotherapeutic methods, came to the conclusion that they "contradict every major premise of behaviorism" (70, p.318); the basic premises of "orthodox" behaviorism being:

> (a) *determinism:* the doctrine that all of man's actions, thoughts, beliefs etc. are ultimately determined by forces outside of his control . . . (b) *epiphenomenalism:* the doctrine that conscious states (e.g. ideas), if they exist at all, are merely by-products of physical events in the body and/or in the external world . . . they have no effect on . . . the individual's subsequent actions or . . . ideas; and (c) the rejection of introspection as a scientific method . . . (71, p.318).

While Wolpe's formal theoretical view is consistent with orthodox behavioristic principles, his treatment procedures are clearly not. Subjects are asked to introspect and so provide data regarding the situations which evoke strong, negative emotions such as anxiety or guilt. The patient's verbal report is used to identify "the content and intensity of his negative emotional reactions" (72, p.321). Such a procedure then focuses upon the patient's conscious experience. Therapy is aimed at modifying phenomenological experience directly, and altering behavior only indirectly as a result of such changes in internal cognitive states.

Another critical feature of Wolpe's procedure which contradicts the canons of behavioristic methodology is the emphasis placed upon the patient's frame of mind upon entering therapy. Every effort is made to assure the client that his problem is treatable; its symptoms reversible. Marcia et al. (73) contend in fact that the success of systematic desensitization procedures in the treatment of phobias, for instance, can largely be accounted for in terms of patient expectancies for improvement. These experimenters presented one group of subjects, a high expectancy group, with mildly painful shocks during exposure of what subjects thought were subliminal presentations of anxiety-evoking stimuli. Subjects were also shown fake polygraph records after each session which indicated gradual improvement. The records thus showed a steady decline in anxiety reactions in response to the slide presentations. Compared to no treatment subjects or subjects who were correctly informed that the slides were in fact empty, and who were given no autonomic feedback, the high expectancy group showed significant anxiety reduction. In fact, the high expectancy group improvement was comparable to that found using traditional desensitization methods. The therapist employing desensitization procedures, as outlined by Wolpe, may then be partially engaged in:

> correcting the patient's wrong ideas and providing him with new knowledge and/or values. This is known in the vernacular as changing the patient's mind! It is not well publicized as a behavioristic technique (74, p.322).

It would appear that "behavioristic" therapies of various types have long been cognitively-oriented, and that the "revolution", to use Mahoney's term, taking place in psychotherapy consists not so much in the fact that "behaviorists and cognitive psychologists . . . are easing into the same theoretical bed" (75, p.5) as in the recognition that they have been unwitting bedfellows for quite some time (76). It is as if behavioristically oriented psychologists themselves have had to be gradually desensitized in order that they would come to recognize the role of cognition in the effectiveness of their procedures. The point of interest for the purposes of this discussion is the fact that such inconsistencies between theory and practice went so long unnoticed. Perhaps such inconsistencies were simply not apparent to traditional behaviorists who thought it possible to separate behavior and cognition; failing to recognize that in intentional movement (action) are reflected aspects of mind itself. Overlooking inconsistencies is of course not peculiar to the behavioristically-oriented. It is a common feature of scientific thinking enamoured of a particular point of view.

Kirsh (77) provides a striking example of the overlooking of inconsistencies in historical data as a consequence of commitment to a paradigm or theoretical view which is widely shared among scientists in a variety of specialties. The theoretical view in question states that:

> the history of science may be seen as a gradual emergence of knowledge and reason out of the darkness of ignorance and superstition . . . It is sometimes portrayed as a struggle between two opposing camps. On the one side stand the forces of darkness: religious dogma, a reliance on theological doctrine as the test of truth, blind faith, and a superstitious belief in supernatural causation of observable phenomena. On the other side are the forces of light: rationalism, empirical observation as the test of truth, skepticism, and a firm commitment to the universality of natural causation (78, p.149).

Kirsh provides evidence that numerous writers have misstated the temporal relations between the witchhunts of an earlier day and the rise of science (79). He suggests that contrary to the popular view "the Middle Ages was not a period of obsessive concern with possession and witchcraft, nor was it a period characterized by official persecution of . . . those suspected of being possessed" (80, p.150). Rather "the growth of demonology and of the witchhunt mania paralleled that of the scientific revolution" (81, p.152). It is during the period between the publication of Copernicus' "Revolution of the Celestial Orbs" (1543) and Newton's Principia (1687) that most historians estimate the witchhunt mania gained strength and peaked:

> During the Middle Ages, the demonological stream lost ground, as scholars became skeptical of belief in extreme demonic power. But from the early Renaissance onward, the demonological stream slowly regained dominance, reaching its culmination in the massive witchhunts of the seventeenth century (82, p.156).

Due to the presumption then that scientific thinking is eminently rational and can but foster the same trends in society, there has been a widespread misinterpretation of certain historical facts, and overlooking of inconsistencies between the model and historical data.

The scientific advances of the twentieth century have done little to erode the preoccupation of the populace with the occult, astrology, and pseudoscience generally. Indeed, it is as if the era of space travel has somehow instigated a renewed fascination with the same. There is and has been then, it would seem, a temporal link between scientific advance and upsurges in superstitious thinking in the general populace.

The claim of Brush (83) that scientists often attach more weight to theoretical or philosophical argument than to empirical evidence should appear more credible having considered the foregoing examples of how models may come to bias and/or restrict conceptual and data analysis. To this point, certain general theoretical biases have been examined; the discussion turns now to a consideration of the influence of the cultural context upon scientific activities.

The Influences of Cultural Milieu upon the Researcher's Interpretations

Scientists, as the reader is well aware, do not operate in a cultural vacuum. From the scientific and nonscientific community alike, the scientist adopts many of the values which specify those problems deemed to be worthy of scientific analysis. The cultural context may blind the experimenter to weaknesses in his model and the data base upon which it rests, particularly if it is generally held that the model is consistent with observation outside of the experimental setting. The psychological literature on sex differences contains some important examples of such difficulties and is discussed next.

Aspects of the Psychological Literature Dealing With Sex Differences

Cultural factors appear to account in large part for the failure to recognize some prevalent methodological errors in design in the research of sex differences. For instance, Rumenik et al. (84) review studies from diverse areas of psychological research which fail to control for sex of the experimenter in cases where this variable is most likely linked to obtained male-female performance differentials. The situation is complex, for the effect of experimenter sex is in part a function of the task involved, age of the subjects etcetera. Yet there is enough evidence to suggest that this variable is substantially affecting results in task performance studies with children and in some types of person perception research. Consider, for example, a study by Pederson et. a.l. (85) which examined the effects of examiner and subject sex on the WISC Arithmetic subtest. Overall performance of both males and females was better when they were tested by female examiners. Female subjects did better when tested by female examiners and also did better than male subjects tested by females. Better performance of male subjects under male examiners did not reach significance. The authors suggested that the disproportionate number of male math teachers and female English teachers could be one factor underlying boys' general superior math ability and girls' superior verbal ability. This difference then, according to the former interpretation, would simply be a function of the socio-cultural context in which learning takes place, rather than being a reflection of a variation in inherent competencies.

Experimenter sex effects have not, for the most part, been dealt with systematically via the design of psychological experiments which could control for their influence. There is a reluctance to acknowledge that "if sex differences affect male-female relations outside of the lab, it is unrealistic to expect such influence suddenly to cease to exist within an experiment" (86, p.874). In part, the failure to consider the influence of such factors as experimenter sex effects and stereotypic male-female roles may be due to the fact that the data generated by such designs often are presumed to be consistent with performance differentials which occur between the sexes outside of the lab situation. For instance, it has long been assumed by most social psychologists that females are more influenceable than males; yet there is little empirical support for such a contention. Eagly (87) reviews numerous studies on persuasion and non-group-conformity which fail to support the hypothesis, as well as major studies in areas such as social learning and conditioning which also are inconsistent with the notion of greater female influenceability:

> Subtracting the percentage of studies reporting significant differences in the male direction from the percentage reporting significant differences in the female direction yields a net difference of only 14% for persuasion and 4% for non-group-conformity studies . . . In group pressure settings . . . significant differences in the female direction climbs to 31% (88, p.95).

It appears that only in group pressure conformity studies is there any substantial degree of evidence for greater female influenceability. Even then it is worthwhile to note that of the 61 group pressure conformity studies reviewed by Eagly, 38 (62%) reported no difference, 21 (34%) found females to be significantly more conforming and 2 (3%) found males to be more conforming (89, p.92). There is evidence which suggests that where greater female influenceability is found, it is importantly linked to the content of the influence induction. In persuasion research there is, as Eagly points out, an established tradition of using messages dealing with social, economic or political issues; topics on which, unfortunately, women are generally less knowledgeable. In this connection, several studies indicate that individuals are more influenceable on topics on which they are less knowledgeable or which are of little interest to them (90, 91). Further, there is recent evidence indicating that persons tend to conform on issues in which their own sex is held to be relatively disinterested and uninformed (92, 93).

Many authors have not considered these methodological flaws, and despite the weight of empirical evidence to the contrary, claim that women are more influenceable in a broad range of situations both within experimental settings and outside of them (94). It is widely held that this greater suggestibility of women is a direct function of traditional passive female social roles (95). Yet as Eagly (96) points out, such an interpretation does not account for the great cross-situational variability in degree of female influenceability in experimental settings, nor does it explain the absence of any significant relation between the degree of influenceability, and how traditional the attitudes of the women in question are. The mechanisms underlying greater female influenceability in group pressure conformity situations hence appear to be much more complex than generally assumed. What is apparent is that because:

> a biased sample of findings suggested that an unvarying pattern of female influenceability was obtained in the laboratory, psychologists attributed the assumed sex difference to inherent differences in yielding tendencies that characterize males and females socialized in our culture (97, p.108).

Eagly offers an alternative explanation in terms of a tendency for women to be more oriented toward interpersonal goals in group settings than are men, and hence more willing to conform in such situations to maintain harmonious group relations (98). This interpretation is, of course, by no means the definitive one. The point of interest here is that the cultural influences operating on the researcher can result in selectivity in data reporting, and insensitivity to particular experimental design flaws. The question of cultural factors in studies dealing with sex differences is then an especially important one.

There has been a gradual shift away from emphasizing the role of the environmental shaping of sex differences toward a view stressing biologically based differences. Findings relating to so-called inherent sex differences in temperament, thinking strategies, and cognitive competencies have been reported (99). Certain contemporary researchers in neuropsychology are suggesting that the brains of men and women may be organized differently. Such differences are held by these investigators to somehow underly women's greater verbal skill and men's superior spatial and mathematical capabilities. Witelson (100), for instance, suggests that the brains of women are less specialized than those of men and that therefore it is easier for women to perform tasks such as reading which require a combination of spatial and linguistic skills. Witelson contends that were such differences in brain organization environmentally induced, they should increase with age. However, female verbal superiority tends instead to decline with age, while male superiority in spatial ability tends to remain constant over the lifespan. According to Witelson, the absence of increments in such differences between the sexes with age implicates biological rather than environmental factors (101). Witelson appears to argue then that in women there is less lateralization of language function to the left hemisphere; and that less lateralization is causally related to superior reading skill.

One of the assumptions underlying such studies is that it is possible in experimental settings to sample linguistic abilities adequately on the basis of some very specialized verbal tasks as are typically included in psychological test batteries e.g. verbal subtest of the WAIS. An additional assumption is that verbal skills are distinct from spatial skills. What is meant by spatial information is, however, a complex issue. Some researchers (102) have suggested that a propositional code for spatial information exists which may at some point be translated into verbal terms as when one draws a map according to a set of verbal rules. At present, it is not possible to determine precisely what type of skill a particular task engages — linguistic, spatial or both. It is therefore difficult to make reasonable claims about where the control processes for such skills are localized in the brain. The problem is similar in certain respects to that posed by Satz's (103) model of developmental dyslexia. Satz contends that young dyslexics suffer primarily from "perceptual" difficulties while older dyslexics have largely overcome such difficulties and are more prone to encounter "verbal comprehension" problems. It is at present, however, not possible to clearly distinguish between perceptual and linguistic problems or to develop tests which assess either skill exclusively. Pattern recognition, for instance, may involve covert verbal labelling. Should this skill be regarded exclusively as a perceptual one or more fundamentally verbal?

This author is not suggesting that differences in brain organization between the sexes may not exist. What is being suggested is that present measures do not adequately define what is meant by "linguistic skill", "spatial skill" and so on. Hence, such measures cannot at present reliably serve as the basis for inferences concerning differences in neural organization for such skills between the sexes. There is, for example, continuing controversy concerning the extent of verbal competence in the right hemisphere in part due to the limitations of current measures of linguistic skill. To attempt to make clear-cut distinctions between spatial and linguistic skills is an endeavour reminiscent of early psychologists' arbitrary distinctions between sensation and perception or perception and memory. It now appears that such parcelling of human cognitive functions was in large part misleading; for these processes appear to be intricately related. For instance, Piaget (104) has demonstrated that what is perceived is in important ways a function of what is remembered and vice versa.

Witelson holds that the differential representation of males and females in various professions may somehow be linked to biological factors; namely differences in neural processing (105). Consider, however, the aforementioned difficulties in making inferences concerning neural organization on the basis of a set of crude, imperfect behavioral measures and tasks. The cultural context in which the researcher works can, it seems, stimulate premature generalizations if these are in accord with prevailing folklore. Also worthy of attention are certain cross-cultural data which suggest that male verbal incompetency may be culture specific. For instance, there is a very low incidence of dyslexia among both males and females in Japan (106). If Japanese script is responsible for the infrequency of reading difficulty in that country, it is significant that both males and females appear to benefit.

Given the paucity of precise information regarding how verbal input is processed neurally and transformed into something meaningful, it would also seem premature to speculate about matters such as which types of script or reading instruction are optimal and for whom. Witelson, however, advances the proposition, based upon alleged differences in neural processing between the sexes, that males and females perhaps ought receive different types of reading instruction or general education.

It will be clear to the reader from the research examples discussed in this chapter that the scientist is no more immune to subjective influences in selecting research problems and interpreting data than are his more nonscientific fellows. Such factors contribute a significant portion of what Rosenthal terms "experimenter effects". According to Rosenthal, "there are things we have learned about human behavior *in spite of* the possible operation of experimenter . . . effects" (107, p.X). In addition, it is here suggested, there are things we have learned about human behaviour precisely because of, rather than in spite of, experimenter effects. We have learned, for instance, that the scientific enterprise involves the scientist's affective as well as his cognitive processes. Psychology as Bischof explains:

> is the science in which the human being is simultaneously subject and object, and can therefore take two "standpoints": the prototype of the human being can be either the other or the investigator himself (108, p.29).

Such a "reflective" psychology would not view experimenter effects strictly as a source of error since such effects themselves provide insights into the subject matter at hand. That experimenter effects are regarded primarily as nuisance factors, something which contaminates data revealing the intricacies of human psychological functioning is an indication of just how nonreflective currently is the science of psychology.

> If we are going to climb up onto platforms and make generalizations about human behaviour, then such generalizations should clearly explain the behavior of climbing up onto platforms and making generalizations about human behaviour (109, p.5).

REFERENCES

1. Rabb, D. Incommensurable Paradigms and Psycho-Metaphysical Explanations. *Inquiry,* 1978, 21, p. 201-212.
2. Ibid, p. 206.
3. Ornstein, R.E. *The Psychology of Consciousness.* New York: Harcourt Brace, Jovanovich Inc., 1977.
4. Ibid.
5. Ibid., p. 33.
6. Ibid.
7. Sperry, R.W. Two Brains Within the Same Head. In Zimbardo, P.G. and Maslach, C. *Psychology for Our Times.* Glenview: Scott, Foresman, and Company, 1977, p. 59-66.
8. Ibid., p. 65.
9. Dewitt, L. Consciousness, Mind and Self: The Implications of Split-Brain Studies. *British Journal for the Philosophy of Science,* 1975, 26, p. 41-46.
10. Puccetti, R. The Mute Self: A Reaction to Dewitt's Alternative Account of the Split-Brain Data. *British Journal for the Philosophy of Science,* 1976, Vol. 27, No. 1, p. 65-73.
11. Sperry, R.W. and Gazzaniga, M.W. Language Following Surgical Disconnection of the Hemispheres. In Darley, F.L. (ed.) *Brain Mechanisms Underlying Speech and Language,* 1967, Proceedings of a conference held at Princeton, Nov., 1965, New York: Grune and Stratton. p. 108-21.
12. Puccetti, R. Brain Bisection and Personal Identity. *British Journal for the Philosophy of Science,* 1973, 24, p. 339-355.
13. Levy, J. Trevarthen, C. and Sperry, R.W. Perception of Bilateral Chimeric Figures Following Hemispheric Disconnection. *Brain,* 1972, 95, p. 61-78.
14. Ibid. p. 67-68.
15. Puccetti, *Brain Bisection and Personal Identity,* p. 339-355.
16. Sperry, R.W. Hemisphere Disconnection and Unity of Conscious Awareness. *American Psychologist,* 1968, 23, p. 723-733.
17. Eccles, J. The Self-Conscious Mind and the Brain. In Popper, K. and Eccles, J. (eds.) *The Self and Its Brain.* New York: Springer International, 1977, p. 225-421.
18. Polanyi, M. Logic and Psychology. *American Psychogist,* 1968, Vol. 12, p. 27-43.
19. Robinson, D. What Sort of Persons are Hemispheres? Another Look At Split-Brain Man. *British Journal for the Philosophy of Science.* 1976, Vol. 27, No. 1, p. 73-79.
20. Sperling, G. A Model for Visual Memory Tasks. *Human Factors,* 1963, 5, p. 19-31.
21. Robinson, *What Sorts of Persons are Hemispheres?* p. 73-79.
22. Taylor, R. Simple Action and Volition. In Brand, M. (ed.) *The Nature of Human*

Action. Glenview: Scott, Foresman and Company, 1970, p. 41-60.

23. Sperry, R.W. Brain Bisection and Mechanisms of Consciousness. In Eccles, J.C. (ed.) *Brain and Conscious Experience*. New York: Springer-Verlag, 1966, p. 298-313.

24. Ibid., p. 304.

25. Finkelman, D. Science and Psychology. *American Journal of Psychology*, June, 1978, Vol. 91, No. 2, p. 179-199.

26. Neisser, U. *Cognitive Psychology*. New York: Appleton-Century Crofts, 1966.

27. Sperry, R.W. *Brain Bisection and Mechanisms of Consciousness*, p. 307.

28. Colby, K.M. Computer simulation of neurotic processes. In Stacey, R.W. and Waxman, B.D. (eds.), *Computers in Biomedical Research*. New York: Academic Press, 1965, p. 491-503.

29. Colby, K.M., Watt, J. and Gilbert, J.P. A computer method of psychotherapy. *Journal of Nervous and Mental Diseases*, 1966, 142, p. 148-52.

30. Thomas, A. Puccetti On Machine Pattern Recognition. *British Journal for the Philosophy of Science*, 1975, Vol. 26, p. 227-239.

31. Gunderson, K. *Mentality and Machines*. Garden City: Anchor Books, 1971.

32. Sutherland, N.S. Computer Simulation of Brain Function. In Brown, S.C. (ed.) *Philosophy of Psychology*. London: Macmillan Press, 1974, p. 259-268.

33. Taube, M. *Computers and Common Sense*. New York: McGraw-Hill, 1961.

34. National Science Foundation. Toward a Model of Brain Function. *Computers and People*, May 1976, p. 12-23.

35. Ibid., p. 15.

36. Puccetti, R. Pattern Recognition in Computers and the Human Brain: With Special Application to Chess Playing Machines. *British Journal for the Philosophy of Science*, 1974, 25, p. 137-154.

37. Ibid.

38. Brand, J. Classification without Identification in Visual Search. *Quarterly Journal of Experimental Psychology*, 1971, 23, p. 178-186.

39. Ingling, N. Categorization: A Mechanism for Rapid Information Processing. *Journal of Experimental Psychology*, 1972, 94, p. 239-243.

40. Posner, M.I. On the Relationship Between Letter Names and Superordinate Categories. *Quarterly Journal of Experimental Psychology*, 1970, 22, p. 279-287.

41. Turvey, M.T. Constructive Theory, Perceptual Systems, and Tacit Knowledge. In Weimer, W.B. and Palermo, D.S. (eds.) *Cognition and the Symbolic Processes*. Hillsdale: Lawrence Erlbaum Associates, 1974, p. 165-180.

42. Puccetti, R. Pattern Recognition in Computers and the Human Brain, p. 149.

43. Ibid., p. 150.

44. Ibid., p. 145.

45. Thomas, A. Puccetti on Machine Pattern Recognition, p. 231.

46. Furst, C. *Origins of the Mind: Mind-Brain Connections.* Englewood Cliffs: Prentice-Hall, New Jersey, 1979.

47. Kolers, P. and Katzman, M.T. Naming sequentially presented letters and words. *Language and Speech,* 1966, 9, p. 84-95.

48. Baron, J. and Thurston, I. An Analysis of the Word-Superiority Effect. *Cognitive Psychology,* 1973, 4, p. 207-228.

49. Davidoff, J.B. *Differences in Visual Perception.* New York: Academic Press, 1975.

50. Ibid.

51. Turing, A.M. Computing Machinery and Intelligence. *Mind,* 1950, 59, p. 433-460.

52. Furst, C., *Origins of the Mind,* p. 18.

53. Sloman, A., *The Computer Revolution in Philosophy: Philosophy, Science and Models of Mind.* Hassocks: Harvester Press Ltd., 1978.

54. Ibid., p. 16.

55. Ibid., p. 19.

56. Ibid., p. 16.

57. Sutherland, *Computer Simulation of Brain Function,* p. 267.

58. Sloman, *Computer Revolution in Philosophy.*

59. Ryle, S., *The Concept of Mind.* New York: Barnes and Noble, 1979.

60. Taube, *Computers and Common Sense.*

61. Sloman, A., *The Computer Revolution in Philosophy.*

62. Ibid., p. 105.

63. Ibid., p. 106.

64. Mahoney, M.J. Reflections on the Cognitive-Learning Trend in Psychotherapy. *American Psychologist,* January, 1977, p. 5-13.

65. Mahoney, M.J. *Cognition and Behavior Modification.* Cambridge: Ballinger Publishing Company, 1974.

66. Meichenbaum, D. *Cognitive-Behavior Modication: An Integrative Approach.* New York: Plenum Press, 1977.

67. Goldfried, M.R. and Davison, G.C. *Clinical Behavior Therapy.* New York: Holt, Rinehart and Winston, 1976.

68. Wolpe, J. The systematic desensitization treatment of neuroses. *Journal of Nervous and Mental Disease,* 1961, 132, p. 189-203.

69. Goldfried and Davison, *Clinical Behavior Therapy.*

70. Locke, E.A. Is "Behavior Therapy" Behavioristic? *Psychological Bulletin,* 1971, Vol. 76, No. 5, p. 318-327.

71. Ibid., p. 318.

72. Ibid., p. 321.

73. Marcia, J.E., Rubin, B.M., and Efran, J.S. Systematic Desensitization: Expectancy Change or Counterconditioning. *Journal of Abnormal Psychology,* 1969, 74, p. 382-387.

74. Locke, *Is "Behavior Therapy" Behavioristic?* p. 318-327.

75. Mahoney, *Reflections on the Cognitive-Learning Trend in Psychotherapy,* p. 5-13.

76. Wilkins, W. Desensitization: Social and Cognitive Factors Underlying the Effectiveness of Wolpe's Procedure. *Psychological Bulletin,* 1971, Vol. 76, No. 5, p. 311-317.

77. Kirsh, I. Demonology and the Rise of Science: An Example of the Misperception of Historical Data. *Journal of the History of the Behavioral Sciences,* 1978, 14, p. 149-157.

78. Ibid., p. 149.

79. Ullman, L.P. and Krasner, L. *A Psychological Approach to Abnormal Behavior.* Englewood Cliffs: Prentice-Hall, 1969.

80. Kirsh, *Demonology and the Rise of Science,* p. 149-157.

81. Ibid., p. 152.

82. Ibid., p. 156.

83. Brush, S.G. Should the History of Science Be Rated X? *Science,* 1974, 183, p. 1164-1172.

84. Rumenik, D.K., Capasso, D.R. and Hendrick, C. Experimenter Sex Effects in Behavioral Research, *Psychological Bulletin,* 1977, Vol. 84, No. 5, p. 852-877.

85. Pedersen, D.M., Shinedling, M.M. and Johnson, D.L. Effects of sex of examiner and subject on children's quantitative test performance. *Journal of Personality and Social Psychology,* 1968, 10, p. 251-254.

86. Rumenik, Capasso and Hendrick, *Experimenter Sex Effects in Behavioral Research,* p. 852-877.

87. Eagly, A.H. Sex Differences in Influenceability. *Psychological Bulletin,* 1978, Vol. 85, No. 1, p. 86-116.

88. Ibid., p. 95.

89. Ibid., p. 92.

90. McGuire, W.J. and Papageorgis, D. The relative efficacy of various types of prior belief-defense in producing immunity against persuasion. *Journal of Abnormal and Social Psychology,* 1961, 62, p. 327-337.

91. Rhine, R.J. and Severance, L.J. Ego-involvement, discrepancy, source credibility, and attitude change. *Journal of Personality and Social Psychology,* 1970, 16, p. 175-90.

92. Goldberg, C. Sex roles, task competence, and conformity. *Journal of Psychology,* 1974, 86, p. 157-164.

93. Goldberg, C. Conformity to majority type as a function of task and acceptance of

sex-related stereotypes. *Journal of Psychology,* 1975, 89, p. 25-37.

94. McGuire, W.J. The Nature of Attitudes and Attitude Change. In Linzey, G. and Aronson, E. (eds.) *Handbook of Social Psychology,* (2nd ed., Vol. 3), Readings: Addison-Wesley, 1966, p. 136-314.

95. Worchell, S. and Cooper, J. *Understanding Social Psychology.* Homewood: Dorsey Press, 1976.

96. Eagly, A.H. *Sex Differences in Influenceability,* p. 86-116.

97. Ibid., p. 108.

98. Ibid.

99. Levison, C.A. Sex Differentiation in Early Infancy: Problems in Methodology and Interpretation of Data. Paper Presented at the meeting of the Midwestern Psychological Association, May 4-6, 1972, Cleveland, Ohio.

100. Witelson, S.F. Sex and the Single Hemisphere: Specialization of the Right Hemisphere for Spatial Processing, *Science,* 1976, Vol. 193, No. 4251, p. 425-427.

101. Ibid.

102. Pylyshyn, Z.W. What the Mind's Eye Tells the Mind's Brain: A Critique of Mental Imagery. *Psychological Bulletin,* 1973, 80, p. 1-24.

103. Satz, P., Rardin, D., and Ross, J. An Evaluation of a Theory of Specific Developmental Dyslexia. *Child Development,* 1971, 42, p. 2009-2021.

104. Piaget, J. and Inhelder, B. *Memory and Intelligence.* New York: Basic Books, 1973.

105. Witelson, S.F. Sex and The Single Hemisphere, p. 425-427.

106. Makita, K. Reading Disability and the Writing System. In Merritt, J.E. (ed.) *New Horizons In Reading:* Proceedings of the Fifth IRA World Congress on Reading. Vienna, Austria, 1974, p. 250-254.

107. Rosenthal, R. *Experimenter Effects in Behavioral Research.* (Enlarged Edition), New York: Irvington Publishers, 1976.

108. Bischof, N. Cited in Brandt, L.W. Scientific Psychology: What For? *Canadian Psychological Review,* 1975, Vol. 16, No. 1, p. 28-34.

109. Bannister, D. Psychology as an exercise in paradox. In Schultz, D.P. (ed.) *The Science of Psychology: Critical Reflections.* Englewood Cliffs: Prentice-Hall, 1970, p. 4-10.

CHAPTER V

COGNITIVE PSYCHOLOGY APPLIED TO THE STUDY OF THE SCIENTIFIC PROCESS ITSELF

The insights provided by cognitive psychology into the nature of science discussed in this chapter by no means represent an exhaustive list. The author's intent is to sketch a few major areas in which cognitive psychology has contributed to an understanding of what is "scientific knowledge" and how it is to be gained.

According to Neisser, the present optimism among psychologists and others regarding the potential contributions of cognitive psychology is premature. He states:

> The study of information processing has momentum and prestige, but it has not yet committed itself to any conception of human nature that could apply beyond the confines of the laboratory. And within the laboratory, its basic assumptions go little further than the computer model to which it owes its existence (1, p.6-7).

Hopefully, the issues discussed in this text will serve as part of the effort to move cognitive psychology beyond the "narrow . . . specialized field" Neisser (2, p.7) holds it is becoming, to one relevant to broader questions of greater epistemological significance. Before discussing the insights provided by cognitive psychology into scientific activities, it is necessary to examine certain alternative models of how science "works"; how it is that scientific discoveries are made; in order to set the framework for the discussion which follows.

THE DYNAMICS OF SCIENTIFIC DISCOVERY

There are numerous alternative conceptions of the mechanisms underlying scientific progress provided by philosophers of science, some of whom are actively engaged in scientific research themselves. A few of these models of how scientific progress is achieved will be discussed here in order to delineate some of the critical points of controversy.

The focus in this discussion is upon the adequacy of various conceptions of how scientific discoveries are made. How it is that the ideas which the scientist develops so often appear to correspond to "reality" and to bear fruit in terms of leading to additional insights. The emphasis here then is upon the issue of why scientific notions so often appear to work, to hold up against the "empirical facts". Polanyi (3, 4) has attempted to provide an answer to this question, often relying on examples from perception research to bolster his arguments. This author is but in partial accord with Polanyi's model. The points of contention to be examined will, it is hoped, open up new possibilities in terms of conceptions of the nature of scientific discovery.

Polanyi, as this author understands him, argues that scientists can discover something about the way the world "really" is, and this skill is based in large part upon sources of knowledge and forms (styles) of thinking of which the scientist is not consciously aware. These sources of implicit knowledge Polanyi terms "tacit knowledge", and these styles of unconscious inference he designates as "tacit inference". Polanyi holds that: ". . . it is impossible to pursue science without believing that it can discover reality . . ." (5, p.27). Polanyi acknowledges, however, that "the bearing that empirical knowledge has upon reality is unspecifiable. There is nothing in any concept that points objectively or automatically to . . . reality" (6, p.61).

To assert that scientific discoveries reveal something of reality is consistent with Polanyi's notion that the scientist somehow intuits aspects of the facts of nature. He experiences flashes of insight, premonitions which lead him to the truth. The essence of Polanyi's position is the view that the scientist must rely upon ". . . an ultimate power of the mind" in order to establish a truth in nature (7, p.29). Polanyi urges that "we . . . be gratified at the capability of *feeling* such subtle, virtually invisible, signs of reality" and that we accept "tacit knowing as a legitimate and, in fact, indispensable source of all empirical knowledge" (8, p.30).

Such a view of how scientific discoveries are made falls into the category of what Medawar terms "romantic conceptions" of science. He describes such a view thus: "In the romantic conception, truth takes shape in the mind of the observer: it is his imaginative grasp of *what might be true* that provides the incentive for finding out, so far as he can, what *is* true" (9, p.54). Such a view leads to an inescapable dilemma — how is it that the scientist comes to grasp what might be true as Polanyi asserts he is capable of doing? It is here suggested, in contrast, that no such intuitions of the truth could result from the scientist imagining what might be true. The notion that the scientist somehow manages to gain insights into reality, and then tests these insights empirically to confirm that they are so is a prevalent one. The cues leading to such an impression of the mechanisms involved in scientific discovery are quite compelling. The primary factor leading to such a view is, it would seem, the fact that scientists *appear* to solve the problems *they set for themselves*. This cue is, however, misleading. Consider the complexities underlying this view. The difficulties arise in specifying how it is that the scientist comes up with such fruitful problems to tackle. Polanyi makes the point as follows:

> It is commonplace that all research must start with a problem . . . But how can one see a problem, any problem, let alone a good and original one? . . . Plato has pointed out . . . in the Meno . . . that to search for the solution to a problem is an absurdity; for either you know what you're looking for, and then there is no problem; or you do not know what you are looking for, and then you cannot expect to find anything (10, p.21-22).

Polanyi solves this problem, as has been discussed, by postulating that somehow the scientist does know what he's looking for. He has ill-defined but nevertheless valuable premonitions about hidden truths (significant problems and their possible solutions). It is in opposition to such a view contended by the present author that scientists do not intuit the truth but rather create, via theoretical models and experiments, what comes to have the appearance of truth for a number of reasons. A quote from Medawar, winner of the Nobel Prize for Medicine in 1960, provides some insights into the basis for this appearance of truth. Medawar states:

> Good scientists study the most important problems *they think they can solve*. It is, after all, their professional business to solve problems, not merely to grapple with them. The spectacle of a scientist locked in combat with the forces of ignorance is not an inspiring one if, in the outcome, the scientist is routed. That is why some of the most important biological problems have not yet appeared on the agenda of practical research (11, p.7).

Considering Medawar's point that scientists tend to tackle only solvable problems, the issue arises as to how they discern in advance which problems are and are not solvable in a given historical period. It is here suggested that scientists do not address problems which nature poses, but that they rather create pseudo-problems in the sense that their solutions are already embedded in the problem. A theory addresses itself to a problem, which is in fact a function of the theory itself, and not "in" the world. The problem is superimposed upon the empirical evidence, rather than logically deduced therefrom. The same is true for the solution to the problem. Experiments designed to test the theory; the adequacy of the proposed solution to a problem, involve artificial *situations* in which the range of possible outcomes has been restricted. The scientist will go on devising these experiments ("situations") until the data gathered fit with his speculations about what the empirical evidence "ought" to be, given his definition of the problem. The solution is then not separable from the problem.

The scientist does not address a problem posed by nature by sifting through whatever the evidence might be. He is instead committed to a particular conception of the problem, and therefore only certain data will do — will be counted as significant and relevant. An example of such a process is provided by the work of the physicist Millikan discussed in a paper by Holton who states:

> Millikan's . . . commitment to an atomistic explanation of electricity predated his experimental verification and indeed helped him to pick his way through initially indifferent data to support his contention (12, p.330).

The scientist then creates solvable problems by devising experiments which serve to provide the data which appear consistent with the theoretical speculations under investigation. Most often the scientist continues to experiment until he can devise such a confirmational situation, rather than dropping his definition of the problem, and so it was with Millikan. It is clear then that experiments are extensions of theories and not independent tests of theory. Scientists do not just happen upon data which confirm their theories. The effort is a concerted and conscious one to devise that situation which provides the data the scientist desires. The most fascinating scientific revolutions, in Kuhn's sense (13), occur when previously solved problems are perceived from a new theoretical stance not to have been solved at all. Usually this entails the creation of new problems which are held to be more promising than the old. Scientific problems held to be solved are hence no more resistant to change than the solutions themselves.

The view which explains Medawar is contrary to the romantic conception of the nature of scientific discovery is that: ". . . truth resides in nature and is to be got at only through the evidence of the senses . . ." (14, p.54). According to this view, intuitions about truth play no significant role in scientific discovery. It is here contended that this view also is somewhat misleading as is Medawar's proposed compromise between the two conceptions. Medawar (15) proposes that intuitions concerning nature's truths are involved when the scientist comes up with an idea, but the scientific process then becomes one of critical analysis of the idea based upon empirical tests (experimentation). He argues that it is erroneous to assume that general laws can be inferred from sensory experience.As was mentioned in the introduction, the rules of inductive and deductive logic apply to sets of statements and do not specify the relation between sets of statements and observations. It is for this reason that Medawar holds that the *origin* of scientific ideas is imagination based on intuitive insights. Reliance upon the senses comes into play with experimentation which

Medawar conceives of as essentially a form of "criticism".

A view of science as essentially a matter of posing questions to be tested against "empirical facts" is satirized in a masterful article by Newell who characterizes it thus:

> Science advances by playing twenty questions with nature. The proper tactic is to frame a general question, hopefully binary, that can be attacked experimentally. Having settled that bits-worth, one can proceed to the next. The policy appears optimal . . . one never risks much, there is feedback from nature at every step, and progress is inevitable. Unfortunately, the questions never seem to be really answered, the strategy does not seem to work (16, p.290).

The strategy does not work in the sense that it is the theorist who specifies whether a problem is valid or not, whether the data suggest the empirical question has been answered or not. These are human cognitive decisions which are in fact arbitrary. Hence it is that, as Newell states, "the questions never seem to be really answered". This is not to imply that experimentation does not lead to useful predictions or have significant practical implications in an applied area. Yet such successes do not obviate the fact that no theory accounts for all the data (17). The possibility remains that the explanations advanced regarding the experimental outcomes are yet flawed; the problem addressed misconstrued in important ways.

Scientific theories take shape via the empirical tests to which they are put. Which tests are required to definitively validate the theory is by no means evident. The empirical tests which are done place the theory, and the scientific problem which underlies it, in a particular context; delimiting and shaping the possible implications drawn from the theory as well as the data gathered. For instance, whether scientists conceived of light as behaving like particles or waves was very much a function of the experiments they designed, for light is of "a unique character . . . in some situations it behaves like waves and in some situations more like particles" (18, p.38).

It should be evident from the preceding discussion that this author is not in agreement with Holton when he states: "It is one of the great advantages of scientific activity that . . . many questions — for example, concerning the 'reality' of scientific knowledge — cannot be asked" (19, p.331). It is suggested that such questions are relevant to the practical activities of scientists in the laboratory and influence their research styles. For instance, a scientist who equates confirmational data with some truth about nature would seem less likely to perceive the possible weaknesses in his experimental design and conceptualizations. Most scientists then contend that their discoveries reveal something of reality rather than that such "truths" are but imaginative constructions. The scientist may hold either that there exists a scientific method which consciously and conscientiously applied will inevitably lead to discoveries, as was the claim of Bacon and Descartes, or he may contend, as did Einstein, that no such method exists and discoveries are simply the product of luck (20). In either case, the assertion that the "discovery" of reality was in fact but an invention of the creative scientific mind is not considered. It is because such a position has not been considered, it is suggested, that we are left with the pseudo-problem which Agassi succinctly states as follows: "Repeatedly science advances and nobody knows why" (21, p.404).

Hooke was an exception among scientists in that he did not hold his scientific theories accorded with reality despite the apparent confirmational data supporting them. In discussing the controversy between Hooke and Newton, Bechler (22) explains that the point of contention was in fact Hooke's rejection of Newton's assertion that the latter's theory did capture something of reality. Hooke, in commenting upon Newton's claim of the necessity of his (Newton's) hypothesis, states:

> In short I will assure him I do as little acquiesce in that (Hooke's theory) for a reality as I do in his (23, p.200).

Hooke's argument was essentially that the hypothesis proposed by Newton was adequate but not certain in that it was not necessary. There were other possible hypotheses which could be advanced to account for the observations other than Newton's mixture hypothesis of light and colours (24). Hooke stated his case thus:

> Tis true I can in my supposition conceive the white or uniform motion of light to be compounded of thousands of compounded motions . . . but I see no necessity of it. If Mr. Newton hath any argument, that he supposeth an absolute demonstration of his theory, I should be very glad to be convinced by it (25, p.114).

What the present author has been arguing is then that scientific advancement is a function of the scientist's creativity and not of his skill in discerning what is. This creativity is evidenced, for example, by the fact that major discoveries in science result from leaps of the imagination often unfettered by previous validated theories. Popper has, for instance, made the point that Newton's celestial dynamics could not have been 'deduced' from Kepler's: "It is only ingenuity which can make this step" (26, p.33). Cohen's analysis of Newton's work, he contends, "reinforces Popper's view of the supreme importance of creative ideas in the advance of science" (27, p.335).

The mechanisms of scientific progress would appear to involve something other than the piecing together of different facets of reality, with each scientist contributing to the gradual accumulation of knowledge by preparing the way for those to come. The suggestion here is, to reiterate, that the scientist creates his own realities. If such be the case, it is sensible then that the scientist should "intuit" what he is about to "discover", since the discovery is not in any simple sense a reflection of reality. Einstein provides a description of the motives of the individual scientist for engaging in research which appears to be consistent in large part with the preceding analysis. He states:

> . . . Man seeks to form for himself, in whatever manner is suitable for him, a simplified and lucid image of the world . . . and so to overcome the world of experience by striving to replace it to some extent by this image.This is what the painter does, and the poet, the speculative philosopher, the natural scientist, each in his own way . . . It is true that this uncertain methodology may in principle give use to many systems of theoretical physics with equal claim, but in fact it has turned out that at any time just one system is generally accepted to be decidedly superior . . . Though there is no logical bridge from experience to the basic principles of theory, in practice it is agreed that the world of experience does define the theoretical system uniquely (28, p.351).

The scientist seeks how to superimpose his theoretical images of the way the world is upon experience. While Einstein concedes that what are confirmational data is not easily discernible since "there is no logical bridge between experience (and) theory"; he nonetheless contends that there is a "harmony" between reality and successful physical theories (29, p.351). As has been mentioned, however, not all theories are successful because they have a superior fit with the data, contrary to Einstein's claim. The decision to rally behind a particular theory often as not stems from the desire to retain certain well-established principles which would be invalidated were the alternative approach accepted. It is often a social decision made by a community of scientists based on the desire for theoretical consistency. Einstein, as most scientists, was in a sense a "conservative" as Holton (30) explains in that he stressed the continuity of physics as in his remark to Carl Seelig:

> With respect to the theory of relativity it is not at all a question of a revolutionary act, but of a natural development of a line which can be pursued through centuries (31, p.362).

Having outlined the basic fallacies, as this author sees them, in various models of the dynamics and significance of scientific discovery, it is appropriate to introduce a discussion concerning the cognitive experience of the scientist. It is a formidable task to examine what are the scientists' tactics in making discoveries since what scientists say they do often is at variance with what is in fact their methodology. Einstein jokingly admonished those who studied the scientist's behavior "not to listen to what a scientist says he does, . . . (but to) look at what in fact he does" (32, p. 401). In the following discussion, concerning the insights which cognitive psychology provides into scientific thinking, the focus is upon why scientists hold on to certain erroneous metaphysical assumptions concerning their research strategies and how these lead to scientific discovery. Also to be examined is how scientific theories come to gain meaning:

> What does the scientist do and how does he behave when he is doing science, or when he says that he is doing science or that he is behaving in a scientific manner? That is a question that is easy to pose but incredibly difficult, even presumptuous, to try to answer" (33, p.340)

Such is the question which it is suggested cognitive psychology may throw some light upon. The goal is to provide a basis for considering this possibility while fully recognizing the complexity of the issue, and hopefully avoiding the "presumptuousness" of which Heibert (34) speaks.

NOTES ON WHY SCIENTISTS SAY THE THINGS THEY DO REGARDING METHODOLOGY

According to Polanyi, science is based on nonexplicit knowing. His views are shared by many renowned scientists, such as Einstein, who speaks of the role of insight and intuition in science. According to Einstein, the discovery of the general laws of physics depends not on a deductive process for "To these elementary laws there leads no logical path, but only intuition, supported by being sympathetically in touch with experience" (35, p.351). There is evidence provided by cognitive psychologists that nonexplicit knowing may in fact affect behavior. It will be argued here, however, that the research also provides a basis for contending that the scientist's frequent claim regarding the role of intuitions of reality in making scientific discoveries is a faulty one.

Turvey has reviewed studies which provide empirical support for the notion of tacit knowledge. For instance, in a study by Wickens (36), words are presented briefly and then masked by a succeeding stimulus. Following the presentations, subjects are instructed to select which one of two words was most similar to the masked (unidentified) word. Half of the time the second word was similar along some dimension to the masked word. It was found that for certain dimensions - semantic differential, taxonomic categories and synonymity - subjects were likely at better than chance to identify the semantically related word. Hence, it appears possible to have "tacit knowledge about the meaning of a word without having explicit knowledge of its identity" (37, p.173). The question arises then as to whether some form of tacit knowledge plays a role in the scientific process. Does the scientist know more than he can tell as Polanyi claims? Are there intuitive experiences which lie at the root of scientific discovery? It is more likely, it is here suggested, that scientists sometimes "tell more than they know", and that their verbal reports of their own mental processes in making discoveries are in certain respects inadequate despite their appeal. This is not to underplay the valuable insights that have been gained into scientific processes via collecting of interview data from scientists themselves (e.g. Mitroff's work). Nisbett and Wilson (38) provide evidence as to how distortion in verbal reporting about mental processes may come about and why such reports yet have an air of plausibility about them.

Several cognitive psychologists such as Neisser (39) have suggested that persons have no introspective direct access to higher mental processes. This is perhaps overstating the case somewhat (see reference note). Nisbett and Wilson cite evidence revealing that people are most often unaware of the effects of stimuli upon the higher mental processes and that:

> When reporting on the effects of stimuli, people may not interrogate a memory of the cognitive processes that operated on the stimuli, instead, they may base their report on implicit, à priori theories about the causal connection between stimulus and response (40, p.233).

What is argued here is that such implicit à priori theories may play a role in the scientist's reports of intuitions about the relation between theoretical perspectives and the significance of certain pieces of empirical evidence.

Nisbett and Wilson report research in a number of different areas which indicate that people are often unaware of the stimuli which affect "higher order, inference - based responses." According to Nisbett and Wilson, much of the evidence demonstrating that persons may not be directly aware of their own cognitive processes comes from the cognitive dissonance literature. This literature demonstrates that persons often make faulty causal attributions in order to account for particular attitudinal, emotional and behavioural responses. For instance, in a study by Zimbardo et. al. subjects were asked to endure a series of painful electric shocks while performing a learning task. At the completion of the task, subjects were requested to endure the series once more. At this point there were two experimental manipulations. Some subjects were given sufficient justification for performing the task a second time by informing them that the success of the entire experiment depended upon their willingness to repeat the task. Another group of subjects were given insufficient justification for repetition of the task and were simply told that the experimenter was curious about what the results might be in a second study. Subjects given insufficient justification for participating in the second study showed lower GSR responses and better learning relative to subjects who had been given sufficient justification. The point of interest for this discussion is that typically in such studies there are no verbal report differences between groups despite the existing behavioural and physiological differences which result from the experimental manipulation.

Hence in the study mentioned, experimental subjects given inadequate justification for continuing with a second shock study did not report that the shock was less painful than did subjects given adequate justification. Yet experimental subjects reacted to the shock in ways which would suggest they did indeed experience less pain, as evidenced for example by the lowered GSR. Subjects often do not seem to be directly aware of how the experimenter's instructions may have modified their cognitive processes or that in fact any significant change has taken place in their responses to particular stimuli. In reviewing the evidence on the type of verbal reports experimental subjects generally give in dissonance and attribution experiments, Nisbett and Wilson state:

> Whatever the inferential process, the experimental method makes it clear that something about the manipulated stimuli produces the differential results. Yet subjects do not refer to these critical stimuli in any way in their reports on their cognitive processes (41, p.239).

It appears to be the case then that often stimuli are registered without conscious awareness despite the profound influence they exert upon our attitudes and behavior. Many more stimuli are registered than can be stored in short or long term memory. In addition, individuals are frequently incorrect when attempting to discern which particular cues led to problem solution as demonstrated by Maier (42). Maier demonstrated that subjects consistently reported that a useless rather than a critical cue, as determined by previous work, had played a role in problem solution on various tasks.

Given the fact that individuals are often poor at discerning which are the stimuli influencing their cognitive processes and how, it is likely that scientists may sometimes be subject to the same difficulties, when offering verbal reports about the steps involved in gathering certain sets of data and arriving at particular conclusions. Scientists then in reporting certain intuitions as to what they might find, and how these intuitions are related to particular observations, may in fact not be providing a completely accurate picture of the actual cognitive processes involved. More direct evidence on this point will be presented shortly. At this point, it is necessary to examine the evidence concerning on just what basis verbal reports, presumed to be a function of introspective data, are frequently made. Nisbett and Wilson contend that such reports are the result of à priori causal theories which provide a plausible explanation for the effect under consideration. These authors suggest then that:

> When subjects were asked about their cognitive processes . . . they did something that may have felt like introspection but which in fact may have been only a simple judgement of the extent to which input was a . . . plausible cause of output (43, p.249).

Consistent with the Nisbett and Wilson view are the findings that observers often give verbal reports about stimulus effects which are as accurate as those given by subjects actually exposed to the stimuli (44). This would be expected if subjects were in fact not relying upon introspective reports, but rather upon à priori causal theories to which observers are equally privy.

What is suggested here is that a failure to appreciate fully the role of imagination in scientific "discovery" underlies the tendency of scientists to generate verbal reports about intuitions of reality which seem eminently reasonable. It is often claimed that such intuitions, of which the scientist becomes cognizant via introspection, are essential to the scientific process. It has been suggested here that such "intuitions" may often be but à priori causal theories held independently of any introspective process.

It would seem to be nonfunctional for the scientist to hold that he intuits something of reality and that the apparent sense of his intuitions are a direct product of the fact that nature is coherent. Yet the belief that the scientist intuits something of reality and that he has introspective access to these intuitions is pervasive. It is this belief perhaps which often keeps scientists from venturing forth with less plausible but perhaps ultimately more fruitful hypotheses. Medawar states that for the scientist there exists: "some internal censorship which restricts hypotheses to those that are not absurd, and the internal circuitry of this process is quite unknown" (45, p.53). The previous discussion will, it is hoped, offer some insights into the "internal circuitry" to which Medawar refers.

Too often scientists sense that their "intuitions" are somehow dictated by their insights into the nature of things and the likelihood of trying out highly novel hypotheses consequently declines. Thus it was with Newton that "some hypotheses which seemed to him too farfetched he refused to adopt . . ." (46, p.251). He was reluctant, for instance, to postulate hypotheses required to account for the causes of gravity for such speculation, he may have thought, could lead only to the most implausible of theories. Were scientists to abandon the notion upheld by Newton that they do not "feign hypotheses" (47), scientific revolutions would perhaps be more frequent and less traumatic than they are at present. Brillouin presents a similar argument when he states: "Theories . . . are human inventions not divine revelations; they will be changed, modified, adapted . . . ad infinitum, as long as scientists keep working." (48, p.45).

The scientist then creates reality via symbols rather than intuiting it. Modern physics has led the scientist far beyond the universe suggested by his senses to a "colourless, soundless, impalpable cosmos which lies like an iceberg beneath the plane of man's perceptions . . . a skeleton structure of symbols" (49, p.114). The limits of what could be, as defined by the nature and structure of the human sensory system, have been surpassed. Evidence from physics research, for instance, has pointed to paradoxes such as electrons that appear to be two places at once. As a result of such apparent paradoxes, it has become evident that "we are always constructing physical reality" (50, p.95).

Pribram contends that studies on the nature of brain processing provide striking additional evidence of this capability to construct an "objective world separated from the receptor surfaces which interface the organism with his environment" (51, p.93). Pribram cites as a case in point a series of studies by Von Békésy (52) in the area of somatosensory perception. Békésy found that a series of vibrators placed on the forearm produce cortical responses consonant with a point perception rather than with the actual pattern of physical stimulation at the receptors. Further, when the vibrators are applied to both forearms, the point perception leaps out into the space between the arms. Here then is evidence of what Pribram terms "projection". Such evidence indicates that not always does "the percept invariably and directly give evidence of the physical organization that gives rise to perception" (53, p.94). We construct our phenomenal world as we do our scientific world for ultimately "Science is a product of the human brain" (54, p.vii).

The significant role of theoretical symbols in the scientists' construction of images of the world has been alluded to. In the next section is examined what cognitive research reveals about the nature of understanding. The question to be addressed is what makes particular theories seem comprehensible.

The Context - Dependent Nature of Knowledge and Understanding

It is the task of the scientist to create a 'framework of logic' to account for empirical data. (55, p.viii). It is this context, it is here suggested, resulting from scientific imagination which makes the data appear comprehensible, and which embues it with particular meanings. Unfortunately, too often it is forgotten that the context was generated by the human mind, and therefore that it limits the possible range of perspectives from which the evidence could conceivably be considered. Kuhn discussed how certain scientific paradigms restrict the scientist's vision and make it difficult, if not impossible, for him to fully appreciate the value of competing paradigms (56).

Cognitive psychologists have provided insights into the context-dependent nature of comprehension. What is understood is very much dependent upon the relationships which are suggested by the context in which the input is embedded. For instance, what Newton perceived to be an anomaly-refracted light rays which produced oblong images - was not so understood by his contemporaries. The latter, perceived the phenomenon to be in accord with the laws of optics (the theoretical and empirical context) of the time (57). Newton's contribution was to create a new mathematically specified context in terms of which the anomaly became apparent and irrefutable.

Much of the work done by cognitive psychologists on the nature of comprehension has been in the area of linguistic comprehension. Many of the findings are, however, relevant to the nature of scientific understanding; of how scientists make sense out of empirical data. The psychological evidence on this topic to be discussed arises from a model of comprehension set out by Bransford and McCarrell (58).Bransford and McCarrell suggest that in understanding, the perceiver makes certain cognitive contributions to the input. These contributions may involve assumptions about "circumstances . . . that may not be immediately perceivable in the environment" (59, p.200). According to these authors, meaning is the result of cognitive contributions and should be conceptualized as "something that is 'created'" (60, p.201). Prior knowledge, according to this model, allows the comprehender to understand the "implicational significancies of events which involve more information than is momentarily present" (61, p.199).

In a series of ingenious studies (62), Bransford and McCarrell demonstrated that in understanding linguistic input the perceiver brings to bear a whole range of assumptions about the way events in the world take place. For instance subjects in recognition tests, involving previously presented lists of sentences plus distractors, invariably falsely recognized foils which described possible pragmatic implications of the situations described in the acquisition sentences. The subjects in processing the acquisition list had made cognitive contributions to it. They had extrapolated from the data given possible pragmatic implications, and could no longer distinguish between the original input and their cognitive contributions to it. Other studies revealed that without the appropriate context, the input as a whole becomes senseless even though its constituents may be familiar and meaningful. For example, a sentence such as "The trip was not delayed because the bottle shattered" is incomprehensible initially since it is not readily apparent what type of situation would allow the "because" relation in the sentence to make sense. Such sentences are extremely difficult to recall and are rated as virtually incomprehensible. Providing a cue such as "christening a ship" eliminates such difficulties since it gives the perceiver the contextual information requisite for understanding.

So too in scientific research elementary empirical observations appear to be incoherent until viewed within the context of an appropriate conceptual framework. As Whewell states:

> To hit upon the right conception is a difficult step, and when the step is once made, the facts assume a different aspect from what they had before . . . possessing new relations before unseen (63).

It is evident that in comprehending language the perceiver brings to bear a network of prior assumptions and extrapolates beyond the data given; drawing innumerable implications of various types; spatial, logical and pragmatic. Such is the case, it is here suggested, also in the interpretation of scientific information. Polanyi (64) makes a similar point in suggesting that scientific theories seem to foster an indeterminate range of implications. He cites as an example the laws of Kepler which "were quite different from anything Copernicus might have expected" but which were nonetheless based on an acceptance of the reality of Copernicus' system (65, p.28). Such examples lead Polanyi to conclude that "whenever we believe in the reality of a thing, we expect that the thing will manifest itself in yet unknown ways in the future" (66, p.27). It is here suggested, in contrast, that human comprehension processes usually involve the creation of possible implications based upon the available data. The drawing of such implications, it would seem, is not fundamentally dependent upon affirmations of the reality of particular concepts or systems of concepts, contrary to Polanyi's claim. Bransford and McCarrell's research demonstrates that persons continually speculate about possible and logical outcomes; forecasting what state of affairs might occur on the basis of what is known.

Polanyi states that the scientific imagination thrusts towards ideas it feels will be confirmed eventually. The scientist's thought, he contends, flows toward particular problems "pointing to a hidden feature of reality" (67, p.40-41). Bransford and McCarrell's research would seem to suggest, in contrast, that the drawing of implications is simply based upon a set of current assumptions about the world, rather than upon intuitions related to "hidden realities". These assumptions then may be erroneous and the implications drawn consequently misleading. It is here contended that scientists ought to accept that the implications they draw from their theories and experiments are human creations not dictated by nature, and that they therefore may be fallacious. On this view, scientists often tackle fruitful problems because of the useful manner in which they define problems and solutions, but the definition of the same is arbitrary. Such a view is in sharp contrast then to Polanyi's who states:

> Most scientists of today view science as an aspect of reality which as such is an inexhaustible source of new and promising problems. Natural science continues to bear fruit because it offers us insight into external reality (68, p.28).

ON DRAWING LOGICALLY NECESSARY CONCLUSIONS FROM SCIENTIFIC THEORIES

Subjective certainty about the veracity of one's intuitions, as is evidenced by the work of Nisbett and Wilson (69), is not an uncommon feature of human thought process. Often, for example, persons are quite certain that particular stimuli have led to problem solution but in fact it can be experimentally demonstrated that these stimuli did not function as clues. Nisbett and Wilson suggest that subjective certainty should be great:

> when causal candidates are salient and highly plausible, but are in reality noninfluential. Subjective certainty should be lower when the causal candidates are actually influential, but are not salient, not plausible, or compete with more salient or plausible but noninfluential causal candidates (70, p.255).

Scientific theories make particular causal candidates appear plausible and salient; that is their function. As a result, there is a sense that one is on the right track, but this of course is not necessarily the case. Polanyi makes a useful point when he states that: "Any questing surmise necessarily seeks its own confirmation" (71, p.41). Yet the sense of commitment that the individual scientist or even a community of scientists has for a particular theoretical view is not to be taken as evidence that the direction of their work is being mapped out by insights into reality; regardless of how compelling such a view may be. Einstein, in responding to Wertheimer's questions regarding the thoughts which led to the theory of relativity, made mention of such a sense of subjective certainty. He stated: "During all those years there was a feeling of direction, of going straight towards something concrete" (72, p.184). Such subjective experiences of certainty may have led Einstein to views similar to those of Polanyi regarding the role of intuition in scientific discovery.

Subjective certainty has both functional and dysfunctional aspects. It allows the scientist to pursue the tedious task of discovery with its endless frustrations and setbacks. However, it may also lead to tunnel vision in that the scientist may come to reject as irrelevant apparent contradictions in the data. This author is in accord with Unger (73) who holds that no one can truly ever be completely certain of anything, and that therefore it is wise to be wary as to just what such feelings of subjective certainty signify. To recognize that knowledge is always context-dependent rather than absolute is to come to more fully appreciate just how tenuous is our grasp of the "facts", regardless of the fruitfulness of particular theories.

In part, the sense of subjective certainty regarding the value of particular scientific theories stems from the notion that they derive logically from what is known empirically. As was mentioned previously, there is no direct link between theory and experience which could allow for classifying the drawing of particular implications from data as logical or illogical. Mach raised similar objections in that "he felt it was imperative for the scientist to recognize the insidious gradual practice by which scientific constructs come to be regarded as philosophically necessary rather than historically contingent" (74, p.341). How are we to know what is a necessary logical derivation from theory or data? Could not the source of the necessity of any necessary statement be but "our having expressly decided to treat that very statement as unassailable?" (75, p.426). If we say we know that certain deductions from theory are necessary because they are in accord with the rules of inference, this hardly solves the problem because it cannot be specified precisely what it means to understand the rules of inference correctly.

Scientists as a community generally reach a consensus of sorts about what particular theories imply. However, as has been discussed, this consensus as to problem definition and implication can hamper creativity. Often the paradigm functions as a basis for:

> group commitments . . . Models supply the group with preferred or permissible analogies and metaphors. By doing so they help to determine what will be accepted as an explanation . . . conversely, they assist in determination of the roster of unsolved

puzzles and in the evaluation of the importance of each (76, p.184).

Scientific progress often results when some of the so-called logical or necessary implications of a theory are abandoned and the theory subsequently leads in new unforeseen directions; that is when new paradigms are invented and the rules of the game change. It is precisely because there is no necessary chain of inferences in scientific theorizing that scientific revolutions are possible. It is further suggested that competing scientific paradigms are incommensurable because they each operate in a sense under different rules of inference, considered by their proponents to lead to logically necessary conclusions.

> The proponents of competing paradigms are always at least slightly at cross-purposes. Neither side will grant all the non-empirical assumptions that the other needs in order to make its case . . . Though each may hope to convert the other to his way of seeing science and its problems; neither may hope to prove his case . . . the proponents of competing paradigms must fail to make complete contact with each other's viewpoints" (77, p.148).

A CASE EXAMPLE: ISSUES IN THEORY VERIFICATION

An example of competing paradigms is provided by the behavioristic versus the cognitive psychological paradigms. The controversy between the two resembles the following hypothetical conversation Wittgenstein imagined might take place between persons who were not playing the same logical game.

> 'But am I not compelled, then, to go the way I do in a chain of inferences? — Compelled? After all I can presumably go as I choose! — 'But if you want to remain in accord with the rules you *must* go this way? — Not at all, I call *this* "accord" (78, p.113-115).

Cognitive theorists such as Brewer argue that "in standard human conditioning paradigms subjects are not making unconscious, automatic responses, but are developing conscious hypotheses and expectations about the experiment, and . . . these produce the resulting 'conditioning'" (79, p.2). Brewer argues that most of the experiments in this field are unable to distinguish between a conditioning theoretical account of the effects and that provided by cognitive theory. In support of the cognitive interpretation of conditioning phenomena, Brewer (80) cites evidence from a number of different types of studies. For instance, several studies have demonstrated that if subjects are simply informed of a CS-UCS pairing, without actually exposing them to such a contingency, conditioning does take place. In fact, subjects only informed of a pairing of shock with a CS showed GSRs of greater amplitude than did control subjects who actually received the pairings. Other studies have demonstrated an absence of conditioning when the CS-UCS relationship is masked by misleading instructions for example.

Studies such as those reviewed by Brewer have been attacked on a number of methodological and conceptual grounds by theorists who favor behavior theory. It has been suggested that the assessment procedure used in studies such as Brewer cites itself often elicits awareness, or that the subject notices his behavior change and only then does he generate an awareness rationale (81, 82). Dulany (83) challenges Brewer's analysis, and that the experimental designs Brewer discusses do not provide definitive evidence in support of cognitive theory as opposed to behavior theory. For instance, the "informed pairing" results (conditioned responses in the absense of CS-UCS pairings when subjects are given expectations about such pairings) are explained by Dulany as due possibly to the experimenter's instructions acting as conditioned stimuli which trigger emotional responses.

Dulany explains that "experimental observation is much too frail and theory-laden" to carry "the evidential weight" necessary to decide which of the opposing frameworks - cognitive or behavioral - is the more adequate (84, p.53). Dulany devalues as a possible solution to such paradigm clashes the rejection of "theoretical credibility in favor of theoretical productivity as the goal of this research" (85, p.52). For Dulany, it is critical that psychologists not abandon realism, he states:

> If we abandon realism, if our aim is not a more *credible* description of cognitive events — not the full story, of course, but a selective abstraction of cognitive reality — what is our claim to a cognitive psychology? (86, p.53).

The earlier discussion in this chapter includes this author's arguments in opposition to assertions regarding realism as the preferred philosophical stance for the scientific researcher. To adopt realism, it is here suggested, leads to extended paradigm clashes which continue beyond the point of theoretical utility. Realism as a philosophical outlook contributes to polarized and unyielding alternative conceptual viewpoints which mitigate more fruitful cooperative efforts. This view then contrasts with that of Dulany, who appears to hold that credible theories are likely to be those which describe reality more accurately. It has been argued here that there is no necessary correlation between the degree of plausibility of a theory, the subjective certainty in its validity it arouses, and the extent of the theory's actual correspondence with reality. Dulany makes an interesting statement which it would appear attests to the weaknesses inherent in claiming that "good" scientific theories describe aspects of reality; he states:

> We all know that, for any array of experimental findings, there is (in principle at least) an unlimited set of theoretical interpretations. In actuality, however, the set of explanations available to the research community is finite and usually rather small. We are constrained by limited imaginations, the known structure of laws, antecedent plausibilities . . . (87, p.54).

It would appear then that realism as an aspect of the scientists' thinking has several disadvantages. Foremost among these disadvantages is that such a perspective tends to restrict the range of theoretical interpretations which are considered in accounting for sets of empirical data. The highly novel theory is looked upon with disfavor; especially if it challenges more established views which have acquired credibility, and are thought to correspond in important ways with what is. Perhaps the problem of "limited imaginations", of which Dulany speaks, stems in large part from the adoption by many scientists of realism as a philosophical position. What, the reader may be asking, is the most viable alternative to realism? The alternative position here suggested is that scientific "knowledge" does not correspond to reality; that it consists of imaginative constructions. It is hoped that such a view, given the discussion to this point, will not be rejected on the grounds that it is too nihilistic, for it is quite the contrary. It is a view which should allow researchers to escape beyond the boundaries set by a limited imagination and a commitment only to that which appears plausible. This argument is carried further in the next section dealing with imagination.

IMAGERY, IMAGINATION AND THE SCIENTIFIC ENTERPRISE

Imagination has long been thought by psychologists to fall outside the domain of phenomena amenable to scientific study. The topic of imagination has not fared well for the most part in the philosophical literature either:

> Ever since Plato, philosophers have condemned recourse to imagery as an inferior form of mental activity — as at best a crutch for, and at worst a debasement of, pure reflection (88, p.x).

Recent research in cognition demonstrates that such views concerning the role of imagery and imagination in problem-solving, and its suitability as an area for scientific investigation are erroneous.

This author has been arguing that scientific theories and inventions are the product of human imagination and that they bear no necessary structural correspondence to reality. Here follows an examination of some of the research on imagery and imagination which it is hoped will substantiate to a degree the position advanced. The discussion which follows then aims to point out that the structure of images in no simple sense can be said to represent what exists external to the imager.

Images: Controversies as to their Structure

Before discussing whether or not images can correspond structurally or functionally with external reality, it is necessary to consider the structure of the image itself. In the psychological literature there can be found those who argue that mental imagery is spatial and modality specific (89), while others hold that the mental representation underlying imagery is one involving abstract propositions (90). Propositions are held to be more abstract than sentences and therefore are not to be equated with that which is necessarily verbal (91). Most imagery theorists do not admit to holding to a picture metaphor for the concept of imagery (92), while it has been argued by others that the picture metaphor is *the* imagery theory (93).

It has been argued that an abstract propositional format provides a medium for the translation or recoding of pictorial information to verbal and vice versa (94). Anderson argues that Pylyshyn's notion of an intermediate code leads to an infinite regress:

> To translate from Code 1 (verbal) to Code 2 (pictorial), it is necessary to translate Code 1 into a new code, Code 3 (abstract propositions) . . . To translate from Code 1 to Code 3 a new Code 4 would be needed and so on (95, p.256).

On the other hand, theorists such as Pylyshyn argue that an intermediate code is necessary to account for how it is that mental images can be interpreted.

It is clear then that there is little agreement among theorists as to the nature of the representations for mental imagery. Anderson holds that "it is not possible for behavioral data to uniquely decide issues of internal representation" (96, p.263). It is also unlikely that physiological data will do so. As Anderson points out, the studies on hemispheric specialization for instance, do not provide direct evidence regarding the form of information representation. Imagery theorists often claim that the fact that the right hemisphere is specialized for spatial tasks and the left for linguistic tasks is consistent with the dual-code notion. However, it could be the case, for example, that:

> rather than having the data differentially stored, . . . the procedures are differentially stored . . . procedures for performing verbal tasks would be in the left hemisphere and procedures for spatial tasks in the right hemisphere. Both types of procedures could take propositional information as their data (97, p.271-272).

It would appear that no theory can distinguish between imaginal and propositional representations on the basis of the data available. This situation damages the theoretical approach which "attempts to discover the 'true' theory" (98, p.270), an approach which this author has been arguing is futile. Anderson (99) holds that though a particular model or theory may not provide a unique explanation of the data, it is yet valuable in that it is capable of being tested. He goes on to make the claim that:

> The fact that the model may be quite indistinguishable scientifically from other quite different models need not be a source of unhappiness . . . If a particular model is equivalent to many other models, we can be more confident in its basic truth (100, p.275).

It would seem that the conclusion that models equivalent to several others must capture some basic truth is far from a necessary one. Anderson himself seems to waffle on this point when he states "Even if the physical implementation directly described in our model proves false of the human brain there are many other ways in which the model could be true" (101, p.275). He does not, however, mention these additional ways in which the model could be "true", and how the fact that it was true could ever be discerned. We are left, in the final analysis, with an indeterminate form of mental representation; perhaps imaginal or propositional or of some alternative type not yet conceptualized. Given the lack of firm evidence regarding the nature of mental representations, it would seem to be premature at best to hold that the structure of human mental representations has some correspondence to external reality. Yet such would seem to be the claim of those who argue in favor of the notion of "structural realism". This claim is examined in the next section.

On Brain Processes, Mental Images, and their Presumed Structural Correspondence to External Reality

According to the great neuropsychologist, Craik, the human mind creates internal models of the external world; that is "physical working models which work in the same way as the processes they parallel" (102, p.51). Craik contended that the organism "carries a 'small scale' model of external reality and of its own possible actions within its head" (103, p.61). The view taken here is in accord with Craik's except in one important respect. It is not claimed here that the internal models constructed by the human brain in fact correspond structurally to external reality. Rather, it is held that these internal models represent imaginative conceptions of what *might* be the case, and that there are no objective criteria available to ascertain whether in fact the models correspond structurally to what *is* the case.

The author's view is in opposition then to what Halwes (104) terms the concept of "structural realism". Halwes gives the following example of the way in which structural realism might explain how persons acquire certain bits of knowledge. Halwes suggests that individuals may come to understand the notion of causality, for instance, because the neural firing in the brain perhaps operates upon such a principle, and thus affords a representation of the principle as it functions in the external world. The problem with this example is that modern day physicists have all but abandoned the principle of causality in interpreting many of their findings. This is not to argue that the principle is not useful for certain purposes; but rather to suggest that the reality of the principle is dubious for "we cannot in any specific situation be certain that what we are observing and measuring reflects reality" (105, p.95). Hence the claim that the brain, if it does embody a representation of the principle of causality, has structural correspondence to reality is also weakened.

Given the indeterminate state of our knowledge of external reality, and of the structure of mental representations and brain processes themselves, it would seem that the claim by Pribram that what we perceive is "a physical universe not much different in basic organization from that of the brain" (106, p.98) does not rest on any firm evidence.

The Role of Imagery in Scientific Discovery

It is this author's view that what is commonly referred to as mental imagery is related in important ways to scientific discovery. The difficulty arises, as has been discussed, in that the actual form of internal mental representations is in doubt, as is any possible link between presumed structural characteristics of such internal models and those of external reality. What is not in doubt is that prominent scientists often give verbal reports of mental imagery experiences which they contend played a crucial role in their arriving at solutions to particular problems. Shepard (107), in an intriguing chapter, cites case histories of many renowned scientists, geometers and others who report such experiences. Shepard argues that the creations of those involved ranging from abstract theories, such as the theory of relativity to concrete inventions, such as the steam engine:

> derived rather directly from nonverbal internal representations or images of a largely spatial and often visual character and that, at least in a metaphorical sense, these creations constitute a very objective and tangible 'externalization' of the corresponding subjective and private images from which they arose (108, p.133).

It is here suggested that such mental representations may sometimes provide the scientist with a clue as to how to structure the data available, as well as a basis for making inferences beyond the information at hand. This need not at all imply that these representations are based upon intuitions of the structure of reality. Yet this is precisely the implication which is all too often drawn. For instance, Kekulé appears to have reached this conclusion when, after having a dream image of a snake swallowing its own tail, he was led to discover the molecular conformation of benzene. He states: "Let us learn to dream gentlemen, and then we may perhaps find the truth" (109, p.135).

In "externalizing" his mental images, to use Shepard's phrase, the scientist is engaged in the process of constructing his physical universe; that is structuring experimental situations which result in particular world views. It is to be remembered that data are situation-specific, and therefore the degree to which they reflect basic features of reality is questionable. Often images permit the scientist to formulate to a degree certain experienced paradoxes. Mach recognized the value of such mental deliberations for the physicist and placed a great deal of emphasis upon "Gedankenexperiments" in the education of his students and in his own research endeavours. The scientist often equates such mental representations with what must be the case externally. This inevitably leads to a sense of commitment to particular imaginative constructions which are pursued until finally realized in some form.

To fully appreciate the subjective elements involved *at every stage* of scientific discovery would, it is here suggested, lead to a rejection of the view that 'The function of science is to discover what *is* the case, not to prescribe what should be the case" (110, p.275). It would seem rather that the scientist's mental representations prescribe for him what should be the case, and that all too often these images, when externalized in the form of theories and models, come to be regarded as accurate descriptions of what is; leaving little room for alternative perspectives. The preceding analysis should lend further support to Kuhn's view that scientific theories can neither be definitively confirmed nor disconfirmed since the link between the scientist's theories and data is but a cognitive one. The presumed structural similarity between the scientist's internal models of the world, and the world itself is then but a metaphysical assumption which, though popular, may in fact turn out to be a barrier to scientific creativity.

REFERENCES

1. Neisser, U. *Cognition and Reality: Principles and Implications of Cognitive Psychology.* San Franscisco. W.H. Freeman and Company, 1976.
2. Ibid, p.7.
3. Polanyi, M. *The Tacit Dimension.* New York: Doubleday and Company, 1966.
4. Polanyi, M. Logic and Psychology. *American Psychologist,* 1968, Vol. 12, p. 27-43.
5. Ibid., p. 27.
6. Polanyi, M. and Prosch, H. *Meaning.* Chicago: University of Chicago Press, 1975.
7. Ibid., p. 29.
8. Ibid., p. 30.
9. Medawar, P.B. Anglo-Saxon Science. *Encounter,* 1965, Vol. 25, No. 2, p. 52-58.
10. Polanyi, *The Tacit Dimension.*
11. Medawar, P.B. *The Art of the Soluable.* London: Methuen and Company, 1967.
12. Holton, G. On the Role of Themata in Scientific Thought. *Science,* 1975, 188, p. 328-334.
13. Kuhn T. *The Structure of Scientific Revolutions.* Second edition, enlarged, Chicago: University of Chicago Press, 1970.
14. Medawar, P.B. *Anglo-Saxon Science,* p. 52-58.
15. Ibid.
16. Newell, A. You Can't Play 20 Questions With Nature and Win. In Chase, W.G. (ed.) *Visual Information Processing.* New York: Academic Press, 1973, p. 283-308.
17. Kuhn, *The Structure of Scientific Revolutions.*
18. Bondi, H. Why Scientists Talk. In Wolfenden, J., Bondi, H., Ashby, E., Beadle, G.W., Gray, J., Adrian, E.D., Esyenck, H.J. and Appelton, E.V. *The Languages*

of Science. New York: Fawcett Books, 1963, p. 35-52.

19. Holton, *On the Role of Themata in Scientific Thought,* p. 328-334.

20. Agassi, J. Scientists as Sleepwalkers. In Elkana, Y. (ed.) *The Interaction Between Science and Philosophy.* Atlantic Highlands: Humanities Press, 1974, p. 391-405.

21. Ibid., p. 404.

22. Bechler, Z. Newton's 1672 Optical Controversies: A Study in The Grammar of Scientific Dissent. In Elkana, Y. (ed.) *The Interaction Between Science and Philosophy,* p. 115-142.

23. Newton, I. Turnball, H.W. (ed.) *Correspondence,* Vol. 1, Cambridge, 1950.

24. Bechler, *Newton's 1672 Optical Controversies,* p. 115-142.

25. Newton, Turnball, (ed.) *Correspondence.*

26. Popper, K. The Aim of Science, *Ratio,* 1957, 1, p.24-35.

27. Cohen, B.I. Newton's Theory vs. Kepler's Theory and Galileo's Theory. In Elkana, Y. (ed.) *The Interaction Between Science and Philosophy,* p. 299-338.

28. Einstein, A., Cited in Elkana, Y. (ed.) *The Interaction Between Science and Philosophy,* p. 351.

29. Ibid.

30. Holton, G. Finding Favor With the Angel of the Lord: Notes Toward the Psychobiographical Study of Scientific Genius. In Elkana, Y. (ed.) *The Interaction Between Science and Philosophy,* p. 349-387.

31. Einstein, A., Cited in Elkana, Y. (ed.) *The Interaction Between Science and Philosophy,* p. 362.

32. Einstein, A. Cited in Agassi, Scientists as Sleepwalkers, p. 401.

33. Heibert, E. Mach's Conception of Thought Experiments in the Natural Sciences. In Elkana, Y. (ed.) *The Interaction Between Science and Philosophy,* p. 339-348.

34. Ibid.

35. Einstein, A., Cited in Holton, *Finding Favor With The Angels of the Lord,* p. 349-387.

36. Wickens, D. Characteristics of word encoding. In Milton, A.W. and Martin, E. (eds.) *Coding Processes in Human Memory.* Washington, D.C.: Winston, 1972.

37. Turvey, M.T. Constructive Theory, Perceptual Systems and Tacit Knowledge. In Weimer, W.B. and Palermo, D.S. (eds.) *Cognition and the Symbolic Processes.* Hillsdale: Lawrence Erlbaum, 1974, p. 165-180.

38. Nisbett, R.E. and Wilson, T. Telling More Than We Can Know: Verbal Reports on Mental Processes. *Psychological Review,* 1977, Vol. 84, No. 3, p. 231-259.

39. Neisser, U. *Cognitive Psychology.* New York: Appleton-Century-Crofts, 1967.

40. Nisbett and Wilson, *Telling More Than We Can Know,* p. 233.

41. Ibid., p. 239.

42. Maier, N.R.F. Reasoning in Humans: II The solution of a Problem and its Appearance in Consciousness. *Journal of Comparative Psychology,* 1931, 12, p. 181-194.

43. Nisbett and Wilson, *Tellng More Than We Can Know,* p. 249.

44. Nisbett, R.E. and Bellows, N. Accuracy and Inaccuracy in verbal reports about influences on evaluations. Unpublished manuscript, University of Michigan, 1976.

45. Medawar, P.B. *Induction and Intuition in Scientific Thought.* Philadelphia, American Philosophical Society, 1969.

46. Elkana, Y. Boltzmann's Scientific Research Program and Its Alternatives. In Elkana, Y. (ed.) *The Interaction Between Science and Philosophy,* p. 243-279.

47. Ibid.

48. Brillouin, L. *Scientific Uncertainty and Information.* New York: Academic Press, 1964.

49. Barnett, L. *The Universe and Dr. Einstein.* New York: Bantum, 1957.

50. Pribram K.H. Some Comments on the Nature of the Perceived Universe. In Shaw, R. and Bransford, J. (eds.) *Perceiving, Acting and Knowing.* Hillsdale: Lawrence Erlbaum Associates, 1977, p. 83-101.

51. Ibid., p.93.

52. Von Békésy, G. *Sensory Inhibition.* Princeton: Princeton University Press, 1967.

53. Pribram, *Some Comments on the Nature of the Perceived Universe,* p. 94.

54. Brillouin, *Scientific Uncertainty and Information,* p. vii.

55. Ibid., p. viii.

56. Kuhn, *The Structure of Scientific Revolutions.*

57. Bechler, *Newton's 1672 Optical Controversies,* p. 115-142.

58. Bransford, J.D. and McCarrell, N.S. A Sketch of a Cognitive Approach to Comprehension: Some Thoughts About Understanding What it Means to Comprehend. In Weimer, W.B. and Palermo, D.S. (eds.) *Cognition and the Symbolic Processes.* Hillsdale: Lawrence Erlbaum, 1974, p. 189-229.

59. Ibid., p. 200.

60. Ibid., p. 201.

61. Ibid., p. 199.

62. Ibid., p. 189-229.

63. Whewell, W. *Philosophy and Discovery.* London: John W. Parker, 1890.

64. Polanyi, *Logic and Psychology,* p. 27-43.

65. Ibid., p. 28.

66. Ibid., p. 27.

67. Ibid., p. 40-41.

68. Ibid., p. 28.

69. Nisbett and Wilson, *Telling More Than We Can Know,* p. 231-259.

70. Ibid., p. 255.

71. Polanyi, *Logic and Psychology,* p. 41.

72. Einstein, A. Cited in Wertheimer, M. *Productive Thinking.* New York: Harper, 1945.

73. Unger, P. *Ignorance: A Case for Skepticism.* Oxford: Clarendon Press, 1975.

74. Heibert, E. Mach's Conception of Thought Experiments in the Natural Sciences. In Elkana, Y. (ed.) *The Interaction Between Science and Philosophy,* p. 339-348.

75. Dummett, M. Wittgenstein's Philosophy of Mathematics. In Pitcher, G. (ed.) *Modern Studies in Philosophy, Wittgenstein: The Philosophical Investigations.* London: Macmillan, 1966, p. 420-447.

76. Kuhn, *The Structure of Scientific Revolutions.* , p. 184.

77. Ibid., p. 148.

78. Wittgenstein, L. *Remarks on the Foundations of Mathematics.* New York: Basil Blackwell Publishers, 1956.

79. Brewer, W.F. There is No Convincing Evidence for Operant or Classical Conditioning in Adult Humans. In Weimer, W.B. and Palermo, D.S. (eds.) *Cognition and the Symbolic Processes,* p. 1-42.

80. Ibid.

81. Greenspoon, J. Reply to Spielberger and Denike Operant Conditioning of Plural Nouns: A Failure to Replicate the Greenspoon Effect. *Psychological Reports,* 1963, 12, p. 29-30.

82. Krasner, L. Verbal Operant Conditioning and Awareness. In Salzinger, K. and Salzinger, S. (eds.) *Research in Verbal Behavior and Some Neurophysiological Implications.* New York: Academic Press, 1967, p. 57-76.

83. Dulany, D.E. On the Support of Cognitive Theory in Opposition to Behavior Theory: A Methodological Problem. In Weimer, W.B. and Palermo, D.S. (eds.) *Cognition and the Symbolic Processes,* p. 43-56.

84. Ibid., p. 53.

85. Ibid., p. 52.

86. Ibid., p. 53.

87. Ibid., p. 54.

88. Casey, E.S. *Imagining: A Phenomenological Study.* Bloomington: Indiana University Press, 1976.

89. Cooper, L.A. and Shepard, R.N. Transformations on Representations of objects in space. In Carteretta, E.C. and Friedman, M. (eds.) *Handbook of Perception: Space and Object Perception,* Vol. 8, New York: Academic Press, (1978).

90. Pylyshyn, Z.W. What the Mind's Eye Tells the Mind's Brain: A Critique of Mental

Imagery. *Psychological Bulletin,* 1973, 80, p. 1-24.

91. Anderson, J.R. Arguments Concerning Representations for Mental Imagery. *Psychological Review,* 1978, Vol. 85, No. 4, p. 249-276.

92. Kosslyn, S.U. and Pomerantz, J.R. Imagery, propositions and the Form of Internal Representations. *Cognitive Psychology,* 1977, 9, p. 52-76.

93. Anderson, *Arguments Concerning Representations for Mental Imagery,* p. 249-276.

94. Pylyshyn, *What the Mind's Eye Tells the Mind's Brain,* p. 1-24.

95. Anderson, *Arguments Concerning Representations for Mental Imagery,* p. 256.

96. Ibid., p. 263.

97. Ibid., p. 271-272.

98. Ibid., p. 270.

99. Ibid.

100. Ibid., p. 275.

101. Ibid.

102. Craik, K.J.W. *The Nature of Explanation.* Cambridge: Cambridge University Press, 1952, p. 51.

103. Ibid., p. 61.

104. Halwes, F. Structural Realism, Coalitions, and the Relationship of Gibsonian, Constructivist, and Buddhist Theories of Perception. In Weimer, W.B. and Palermo, D.S. (eds.) *Cognition and the Symbolic Processes,* p. 367-383.

105. Pribram, K.H. *Some Comments on the Nature of the Perceived Universe,* p. 95.

106. Ibid., p. 98.

107. Shepard, R.N. Externalization of Mental Images and the Act of Creation. In Randhawa, B.S. and Coffman, W.E. (eds.) *Visual Learning, Thinking and Communication.* New York: Academic Press, 1978, p. 133-189.

108. Ibid., p. 133.

109. Kekulé, F.A., Cited in Mackenzie, N. *Dreams and Dreaming,* London: Aldus Books, 1965.

110. Anderson, J.R. *Arguments Concerning Representations for Mental Imagery,* p. 275.

NOTES

1. There is continuing controversy as to just how much direct access the individual does in fact have to his higher mental processes. Ericsson and Simon argue that verbal reports of cognitive process are often more accurate than Nisbett and Wilson claim. They contend that inaccurate or incomplete verbal reports occur in studies such as cited by Nisbett and Wilson due to, for instance, problems in short-term memory which are in large part an artifact of the experimental procedure used. It seems reasonable, however, to suggest that much of the evidence can be interpreted as supporting the notion that introspective verbal reports are not immune to distortion, and that such reports may not always reflect accurately upon actual cognitive process.

Ericsson, K.A. and Simon, H.A.; Verbal Reports as Data. *Psychological Review,* 1980, Vol. 87, No. 3, p. 215-251.

CONCLUDING REMARKS

This book has presented a particular view of the psychology of the scientist and the role it plays in scientific process. On this view, subjectivity enters into *every* aspect of scientific activity, including theory justification via empirical test. It has been noted that it appears more difficult frequently for the psychologist to admit of such subjectivity than for scientists in other fields. This perhaps because "early psychology had to free itself from the dominance of philosophy . . . which at the that time was rationalistic and abstract" (1, p.26). Those most likely to incorrectly regard the author's position as "anti-science" are then perhaps psychologists with a more pronounced positivist bias.

It is not the intent of this author to engage in what has been termed psychologism: "the reduction of every matter and concern of science to psychology" (2, p.270). Rather, the goal has been to suggest that psychology can and has provided useful insights into the cognitive life of the scientist. Often as not these insights have been overlooked or rejected out of hand as they tend to challenge current idealized characterizations of scientists and science. Neither is it the intent of this author to attempt to blur the distinction between science and nonscience. This point requires further elaboration. As Kuhn has explained, many pseudo-scientific theories are testable and potentially falsifiable (3). Scientific theories are often not immediately testable and sometimes not completely so. Further, the "fit" between theory and data would appear to be as contrived and invented for scientific models as it is for nonscientific ones, as has been discussed. The correspondence between theoretical implication and empirical evidence then is dependent upon the definition of the problem, which may always conceivably be otherwise, the selection of data and so on. Attempts to draw demarcation lines between science and nonscience via an appeal to empirical reducibility: "far from defeating the supposed enemy metaphysics, in effect presented the enemy with the keys to the beleaguered city" (4, p.254). Rationality, as manifest in the logic underlying experimental design, for instance, also seems inadequate to distinguish that which is scientific. As has been pointed out that logic is far from flawless.

Science, it seems, is an "experimental philosophy" in that it involves the framing of philosophical arguments in "empirical" terms. The "empirical" situation is but a restatement of the philosophical position itself, and its underlying assumptions. There is then no necessary or logical relation between theory and data; the link is a cognitive one depending upon incisive argument. Whether any piece of "evidence" clearly confirms or discredits a certain model is a function of the (data) interpretations available, and is criteria dependent:

> It is a mistake to believe that nature cannot lie, and that, if you listen with an unbiased mind to what nature has to tell you in response to your questioning experiments . . . then you will have a truth (5, p.143).

Neither confirmational nor disconfirmational verification strategies would seem to be equal to the task of securing that truth to which Eccles makes reference.

Though scientists often, not always, reject such a view of science as outlined above, in practice they tend to behave in a manner consistent with such a perspective. This is particularly so when scientific models are being revised drastically. "Paradigm shifts", to use Kuhn's term, are possible in that the scientist recognizes that the discarded model represents a certain *psychological* reality reflected in the particular *arbitrary* definition of the problem the originators of the model entertained. At such points, the scientist becomes for a time self-reflective. In contrast, the pseudoscientist carries his philosophical "realism" always into practice and relinquishes it not for a moment. His assumptions are held to derive from observation (nature) and thus are finite, and unassailable. Denied is the critical role of the psychology of the theorist in the articulation of problems and in data interpretation. In contrast, while the scientist may sometimes initially reject anomalous data and the competing views which brought it to light, ultimately such data is dealt with. The fallibility of established views and the apparently confirmational data supporting them is reluctantly acknowledged.

This author's work can best be regarded as, in part, an invitation to scientists to be more self-reflective also during periods of "normal science" (6). Periods of normal science being those in which science progresses via an accumulation of data based on well-established frameworks and previous findings. Such a reflective normal science would, it is suggested, lead to more objective, critical evaluation of the existing scientific wisdom (dogma), and promote creative shifts in scientific outlooks. The author would hope that this work has contributed to the understanding that:

> We can no longer afford to ignore psychology in any future accounts of science . . . [and] that one of the greatest purposes of [science] is as much — perhaps even more so — to learn about ourselves as it is to learn about nature (7, p.270).

REFERENCES

1. Van Kaam, A. Assumptions in psychology. In Schultz, D.P. (ed.) *The Science of Psychology: Critical Reflections.* Englewood Cliffs: Prentice-Hall, 1970, p. 24-29.
2. Mitroff, I. *The Subjective Side of Science: A Philosophical Inquiry into the Psychology of the Apollo Moon Scientists.* Amsterdam: Elsevier, 1974.
3. Kuhn, T. Logic of discovery or psychology of research? In Lakatos, I. and Musgrave, A. (eds.) *Criticism and the Growth of Knowledge.* p. 1-23.
4. Popper, K. *Conjectures and Refutations.* New York: Basic Books, 1962.
5. Eccles, J. *Facing Reality: Philosophical Adventures of a Brain Scientist.* London: Longmans, 1970.
6. Kuhn, T. *The Structure of Scientific Revolutions.* (second edition, enlarged) Chicago: University of Chicago Press, 1970.
7. Mitroff, *The Subjective Side of Science.*

AUTHOR INDEX

About The Author

Sonja C. Grover is also author of *A Manual for the Analysis of Human Behavior in the Psychological Laboratory* (in press). Before attending graduate school, she spent a year at the Psychology Department, University of Sussex, England, working as a research assistant. She received her M.A. from Lakehead University, Ontario, in 1973 and Ph.D. from the University of Toronto in 1976. After receiving her doctorate, she was an educational researcher with University of Toronto while also teaching part-time at Nipissing University. She has been Assistant Professor at the Faculty of Education, Queen's University, Ontario, and at the Department of Psychology, Lakehead University. Subsequently she assumed the position of senior psychologist with Alberta Mental Health and taught a course in psychoeducational assessment at the University of Alberta. Currently she is on the faculty at the Department of Educational Psychology at the University of Calgary, Alberta. She has published articles in various areas including language disorder, inferential thought processes, word recognition, and moral development. She has just completed a book entitled *The Cognitive Basis of the Intellect: A Response to Jensen's "Bias in Mental Testing"* and is collaborating on a book for educators dealing with cognitive process.